Applied Welding Engineering: Processes, Codes and Standards

Applied Welding Engineering: Processes, Codes and Standards

By Ramesh Singh

ELSEVIER

AMSTERDAM • BOSTON • HEIDELBERG • LONDON
NEW YORK • OXFORD • PARIS • SAN DIEGO
SAN FRANCISCO • SINGAPORE • SYDNEY • TOKYO

Butterworth-Heinemann is an imprint of Elsevier

Butterworth-Heinemann is an imprint of Elsevier
225 Wyman Street, Waltham, MA 02451, USA
The Boulevard, Langford Lane, Kidlington, Oxford, OX5 1GB, UK

First edition 2012

Notice
No responsibility is assumed by the publisher for any injury and/or damage to persons or
property as a matter of products liability, negligence or otherwise, or from any use or operation
of any methods, products, instructions or ideas contained in the material herein. Because of rapid
advances in the medical sciences, in particular, independent verification of diagnoses and drug
dosages should be made

British Library Cataloguing-in-Publication Data
A catalogue record for this book is available from the British Library

Library of Congress Cataloging-in-Publication Data
A catalog record for this book is available from the Library of Congress

ISBN: 978-0-12-391916-8

For information on all Butterworth-Heinemann publications
visit our Web site at www.elsevierdirect.com

Typeset by MPS Limited, a Macmillan Company, Chennai, India
www.macmillansolutions.com

Printed and bound in the United States of America

11 12 13 14 15 16   10 9 8 7 6 5 4 3 2 1

**Working together to grow
libraries in developing countries**

www.elsevier.com | www.bookaid.org | www.sabre.org

ELSEVIER   BOOK AID
International   Sabre Foundation

This book is dedicated to the memory of Sgt SA Siddique of the Indian Air Force. I am eternally grateful to Sgt Siddique for instilling the seeds of metallurgy and welding engineering in my mind, and for his training in thought processing and hard work. This book is just a small token of gratitude to the great teacher.

Contents

Contents

Section 4
Codes and Standards

There are several books on the market that address the needs of academia. Some others address specific topics, and are aimed at that particular segment of readers who are aware of the subject matter but are in search of new perspectives or new findings.

This book, Applied Welding Engineering, aims to bridge the gap left by the two segments described above. It intends to support the under-supported, by giving a practical perspective to the theoretical texts. Hopefully the students trying to steer through the terminologies of the field, balancing theory with the practical side of welding engineering, will find this book useful in bridging that gap. The objective is to keep the budding engineers moored in the theory taught in university and colleges while exploring the real world of practical welding engineering. The book is also aimed at engineers, non-engineers, managers and inspectors, to serve as a body of knowledge and source of reference.

In writing this book I do not claim originality on all thoughts and words; on a universal subject like welding engineering no single source can claim the originality of thought. A lot of information contained in this book comes from my personal experience, and also from several industry publications like the American Society of Mechanical Engineers (www.asme.org), American Welding Society (www.aws.org), American Society of Metals (www.asminternational.org), NACE International (www.nace.org), American Petroleum Institute (www.api.org), etc. and several training manuals including The Welding Institute, UK (www.twi.co.uk), Indian Air Force training manuals, ASNT (www.asnt.org), the Canadian Standard Association (www.cas.com) and Canadian General Standard Board (CGSB) (www.tpsgc-pwgsc.gc.ca), just to name a few. It is not possible for me to distinguish which part of my experience is gained from which specific source, but I cannot deny their combined contribution in developing my knowledge base over the years. I acknowledg them all, and I am proud of that. Where I have consciously borrowed matters and ideas directly from theses sources, I have acknowledged them as best as I can and appreciate the great service these bodies have rendered to welding engineering.

Those individuals who need more detailed study on any specific topic covered in this book must reach out to these specialized associations and institutions for further guidance. There are several published works available from these bodies that can be of help in developing in-depth understanding of specific subjects.

Preface

Acknowledgment

Writing this book made me realize how dependent a person is on others in accomplishing a task of this nature. The process started with retrieving several years of notes, handouts and hand-written chits. Some were from as far back as 1969, some of them had turned yellow and were torn, stained with sweat and dirt, possibly from those physical punishments that were liberally given to us as students in the Air Force Institute. Some of the papers were torn at the folds, and tapes were used to save them until I had used the information contained there. I needed help for all this.

I am extremely grateful to the management and team of Gulf Interstate Engineering, Houston (www.gie.com) for creating an environment that encouraged me to write this book. I am especially grateful to James McLane Jr. III, my friend and colleague at Gulf Interstate Engineering, who encouraged me to take up this project. I am also indebted to the encouragement, support and help from my friend Olga Ostrovsky. She helped me to negotiate the obstacles of writing and editing the drafts. Without her expert help this book would not have been possible.

Last, but not the least, I am also grateful to my loving wife Mithilesh, and my son Sitanshu for their support in accomplishing this goal. Mithilesh tolerated my indulgence with the project. Without her support and understanding this task would not have been possible.

Finally, a few words on the dedication of this book. I have dedicated this book to Sergeant SA Siddique of the Indian Air Force. Sergeant Siddique taught me the first lessons of metallurgy and welding engineering. Drawing on my Indian ethos I know the protocol, "Teacher takes precedence even over God", hence the dedication.

Section 1

Introduction to Basic Metallurgy

Introduction

When we talk of metallurgy as being a science of metals, the first question that arises in the mind is what is a metal?

Metals are best described by their properties. They are crystalline in the solid state. Except for mercury, metals are solid at room temperature; mercury is a metal but in liquid form at room temperature. Metals are good conductors of heat and electricity, and they usually have comparatively high density. Most metals are ductile, a property that allows them to be shaped and changed permanently without breaking by the application of relatively high forces. Metals can be either elements, or alloys created by man in pursuit of specific properties. Aluminum, iron, copper, gold and silver are examples of metals which are elements, whereas brass, steel, bronze etc. are examples of manmade alloy metals.

Metallurgy is the science and technology of metals and alloys. The study of metallurgy can be divided into three general sections.

1. Process metallurgy
    Process metallurgy is concerned with the extraction of metals from their ores and the refining of metals. A brief discussion on the production of steel, castings and aluminum is included in this section.
2. Physical metallurgy
    Physical metallurgy is concerned with the physical and mechanical properties of metals as affected by their composition, processing and environmental conditions. A number of chapters in this section specifically address this topic.
3. Mechanical metallurgy
    Mechanical metallurgy is concerned with the response of metals to applied forces. This is addressed in subsequent chapters of this section.

PURE METALS AND ALLOYS

Pure metals are soft and weak and are used only for specialty purposes such as laboratory research work, or electroplating. Foreign elements (metallic or non-metallic) that are always present in any metal may be beneficial, detrimental or have no influence on a particular property. Disadvantageous foreign elements are called impurities, while advantageous foreign elements are called alloying elements. When these are added deliberately, the resulting metal is called an alloy. Alloys are grouped and identified by their primary metal element, e.g. aluminum alloy, iron alloy, copper alloy, nickel alloy etc.

Most of the metallic elements are not found in a usable form in nature. They are generally found in their various oxide forms, called ores. Metals are recovered from these ores by thermal and chemical reactions. We shall briefly discuss some of these processes. Those for the most common and most abundantly used metal – iron – are discussed in the following paragraphs.

SMELTING

Smelting is an energy-intensive process used to refine an ore into a usable metal. Most ore deposits contain metals in the reacted or combined form. Magnetite (Fe_3O_4), hematite (Fe_2O_3), goethite ($\alpha FeO(OH)$), limonite (generic formula: $FeO(OH).nH_2O$) and siderite ($FeCO_3$) are iron ores, and Cu_5FeSO_4 is a copper ore. The smelting process melts the ore, usually for a chemical change to separate the metal, thereby reducing the one to metal or refining it to metal. The smelting process requires lots of energy to extract the metal from the other elements.

There are other methods of extraction of pure metals from their ores: application of heat, leaching in a strong acidic or alkaline solution, and electrolytic processes are all used.

IRON

The modern production process for recovery of iron from ore includes the use of blast furnaces to produce pig iron, which contains carbon, silicon, manganese, sulfur, phosphorus, and many other elements and impurities. Unlike wrought iron, pig iron is hard and brittle and cannot be hammered into a desired shape. Pig iron is the basis of the majority of steel production.

Sponge Iron

Removing the oxygen from the ore by a natural process produces a relatively small percentage of the world's steel. This natural process uses less energy and is a natural chemical reaction method. The process involves heating naturally occurring iron oxide in the presence of carbon, which produces 'sponge iron'. In this process the oxygen is removed without melting the ore.

Iron oxide ores, as extracted from the earth, are allowed to absorb carbon by a reduction process. In this natural reduction reaction, as the iron ore is heated with carbon it gives the iron a pock-marked surface, hence the name sponge iron. The commercial process is a solid solution reduction; also called direct-reduced iron (DRI). In this process the iron ore lumps, pellets, or fines are heated in a furnace at 800–1,500°C (1,470– 2,730°F) in a carburizing environment. A reducing gas produced by natural gas or coal, and a mixture of hydrogen and carbon monoxide gas provides the carburizing environment.

The resulting sponge iron is hammered into shapes to produce wrought iron. The conventional integrated steel plants of less than one million tons annual capacity are generally not economically viable, but some of the smaller capacity steel plants use sponge iron as charge to convert iron into steel. Since the reduction process is not energy intensive, the steel mills find it a more environmentally acceptable process. The process also tends to reduce the cost of steel making. The negative aspect of the process is that it is slow and does not support large-scale steel production.

Iron alloys that contain 0.1% to 2% carbon are designated as steels. Iron alloys with greater than 2% carbon are called cast irons.

Alloys

ALLOYS

An alloy is a substance that has metallic properties and is composed of two or more chemical elements, of which at least one, the primary one, is a metal. A binary alloy system is a group of alloys that can be formed by two elements combined in all possible proportions.

Homogeneous alloys consist of a single phase and mixtures consist of several phases. A phase is anything that is homogeneous and physically distinct if viewed under a microscope. When an allotropic metal undergoes a change in crystal structure, it undergoes a phase change.

There are three possible phases in the solid state:

- Pure metal
- Intermediate alloy phase or compound
- Solid solution.

Compounds have their own characteristic physical, mechanical, and chemical properties and exhibit definite melting and freezing points. Intermetallic compounds are formed between dissimilar metals by chemical valence rules, and generally have non-metallic properties; Mg_2Sn and Cu_2Se are examples of these.

Interstitial compounds are formed between transition metals such as titanium and iron with hydrogen, oxygen, carbon, boron, and nitrogen. They are usually metallic, with high melting points and are extremely hard; TiC and Fe_3C are examples of interstitial compounds.

Electron compounds are formed from materials with similar lattice systems and have a definite ratio of valence electrons to atoms; Cu_3Si and FeZn are examples of electron compounds.

Solid solutions are solutions in the solid state and consist of two kinds of atoms combined in one kind of space lattice. The solute atoms can be present in either a substitutional or an interstitial position in the crystal lattice.

There are three possible conditions for solid solutions:

- Unsaturated
- Saturated and
- Supersaturated.

The solute is usually more soluble in the liquid state than in the solid state. Solid solutions show a wide range of chemistry so they are not expressed as a chemical formula. Most solid solutions solidify over a temperature range, rather than having a defined freezing point.

Having gained this basic understanding of alloy formation and type of alloy, we move forward to learn about a specific alloy – steel – and the effects of various alloying elements on its properties.

EFFECTS OF ALLOYING ELEMENTS

Carbon Steels

Metals are alloyed for a specific purpose, generally with the aim of improving a property or a specific set of properties. In order to take full advantages of such alloying, it is important that the resulting property of alloying elements is known. In the following discussions we shall learn, with the help of steel metallurgy, about some of the most common alloying practices and the resulting alloy metals.

Sulfur

Sulfur in steel is generally kept below 0.05% as it combines with iron to form FeS, which melts at low temperatures and tends to concentrate at grain boundaries. At elevated temperatures, high sulfur steel becomes hot-short due to melting of the FeS eutectic. In free-machining steels, the sulfur content is increased to 0.08% or 0.35%. The sulfide inclusions act as chip breakers, reducing tool wear.

Manganese

Manganese is present in all commercial carbon steels in the range of 0.03% to 1.00%. Manganese functions to counteract the effect of sulfur by forming

MnS. Any excess manganese combines with carbon to form Mn_3C; the compound associated with cementite. Manganese also acts as a deoxidizer in the steel melt.

Phosphorus

Phosphorus in steel is kept below 0.04%. The presence of phosphorus at levels over 0.04% reduces the steel's ductility, resulting in cold-shortness. Higher levels (from 0.07% to 0.12%) are included in steels that are specifically developed for machining, to improve cutting properties.

Silicon

Silicon is present in most steels in the 0.05% to 0.3% range. Silicon dissolves in ferrite, increasing its strength while maintaining ductility. Silicon promotes deoxidation in the molten steel through the formation of SiO_2, hence it is an especially important addition in castings.

ALLOY STEELS

Plain carbon steel is satisfactory where strength and other property requirements are not severe, and when high temperatures and corrosive environments are not a major factor in the selection of a material. Alloy steels have characteristic properties, due to some element other than carbon being added to them. Alloying elements are added to obtain several properties including the following:

- Increased hardenability
- Improved strength at ambient temperatures
- Improved mechanical properties at low and high temperatures
- Improved toughness
- Improved wear resistance
- Increased corrosion resistance
- Improved magnetic permeability or magnetic retentivity.

There are two ways in which alloyed elements are distributed in the main constituents of steel:

- Dissolved in ferrite
- Combined with carbon to form simple or complex carbides.

THE EFFECT OF ALLOYING ELEMENTS ON FERRITE

Nickel, aluminum, silicon, copper, and cobalt are all elements which largely dissolve in ferrite. They tend to increase the ferrite's strength by solid solution hardening.

Alloying elements change the critical temperature range, eutectoid point position, and location of the alpha (α) and gamma (γ) fields on the iron-iron carbide phase diagram. These changes affect the heat-treating requirements and final properties of alloys.

EFFECTS OF ALLOYING ELEMENTS ON CARBIDE

Carbide-forming elements, including manganese, chromium, tungsten, molybdenum, vanadium, and titanium, increase room temperature tensile properties since all carbides are hard and brittle. The order of increasing effectiveness is chromium, tungsten, vanadium, molybdenum, manganese, nickel, and silicon. Of these, nickel and silicon do not form carbides. Complex carbides are sluggish and hard to dissolve. They act as inhibitors to grain growth and often improve high temperature properties. Chromium and vanadium carbides are exceptionally hard and wear resistant.

Tempering temperatures are raised significantly and in some cases secondary hardening may occur with higher tempering temperatures due to the delayed precipitation of fine alloy carbides.

Some of the alloys in general use are discussed briefly here.

Nickel Steels (2xx Series)

Nickel has unlimited solubility in γ-iron and is highly soluble in ferrite. It widens the range for successful heat treatment, retards the decomposition of austenite, and does not form carbides.

Nickel promotes the formation of very fine and tough pearlite at lower carbon contents, thus, toughness, plasticity, and fatigue resistance are improved. Nickel alloys are used for high-strength structural steels in the as-rolled condition and for large forgings that cannot be hardened by heat treatment.

Nickel-Chromium Steels (3xx Series)

The effect of nickel on increasing toughness and ductility is combined with the effect of chromium on improving hardenability and wear resistance. The combined effect of these two alloying elements is often greater than the sum of their individual effects.

Manganese Steels (31x Series)

Manganese is one of the least expensive of the alloying elements and is always present as a deoxidizer and to reduce hot-shortness. When the manganese content exceeds 0.8%, it acts as an alloying element to increase strength and hardness in high carbon steels. Fine-grained manganese steels have excellent toughness and strength.

Steels with greater than 10% manganese remain austenitic after cooling and are known as Hadfield manganese steel. After heat treatment, this steel has excellent toughness and wear resistance as well as high strength and ductility. Work hardening occurs as the austenite is strain hardened to martensite.

Molybdenum Steels (4xx Series)

Molybdenum has limited solubility in α and γ-iron and is a strong carbide former. It has a strong effect on hardenability and increases high-temperature strength and hardness. Molybdenum alloys are less susceptible to temper brittleness. Chromium-molybdenum alloys (AISI 41xx) are relatively cheap and ductile, have good hardenability, and are weldable.

Chromium Steels (5xx Series)

Chromium forms both simple (Cr_7C_3 and Cr_4C) and complex carbides [$(FeCr)_3C$]. These carbides have high hardness and resist wear. Chromium is soluble up to about 13% in γ-iron and has unlimited solubility in ferrite. It increases the strength and toughness of the ferrite and improves high-temperature properties and corrosion resistance.

Physical Metallurgy

In the previous chapters we briefly introduced the alloying process and various alloys of steel, and so in this chapter we turn to the fundamental physics of metallurgy.

In the solid state materials have a crystal structure, broadly defined as the arrangement of atoms or molecules. The arrangement at the atomic and molecular level is collectively called the microstructure of material. The arrangement may include the abnormalities in the crystalline structure.

CRYSTAL LATTICES

The three-dimensional network of imaginary lines connecting the atoms in a regular solid structure is called the space lattice. A crystal is an arrangement in three dimensions of atoms or molecules in a repetitive pattern. The smallest unit that possesses the full symmetry of the crystal is called the unit cell, the edges of which form three axes, a, b and c. The three-dimensional aggregation of unit cells in the crystal forms a space lattice, or Bravais lattice. Seven different systems of crystal are known, each with a different set of axes and unit cell edge lengths corresponding to the three axes. The corresponding faces of the unit cell are identified with capital letters A, B and C. The interaxial angles α, β and γ correspond to the faces A, B and C. The interaxial angles and edge lengths of the unit cell are unique to each crystalline substance.

CRYSTAL STRUCTURE NOMENCLATURE

The seven crystal structures listed below in the left column are the basic systems. When they are associated with the five lattices, the combination generates 14 different combinations, shown in the right column. The basis for this combination is William B Pearson's widely used Pearson symbols. The system uses lower case letters to identify the crystal system and upper case letters to identify space lattices. The Pearson symbols are given in the brackets.

Triclinic also called anorthic	Primitive	(aP)
Monoclinic	Primitive	(mP)
	Base centered	(mC)
Orthorhombic	Primitive	(oP)
Base centered		(oC)
Body centered		(oI)
Face centered		(oF)
Tetragonal	Primitive	(tP)
Base centered		(tI)
Hexagonal	Primitive	(hP)
Rhombohedral*	Primitive	(hR)
Cubic	Primitive	(cP)
Body centered		(cI)
Face centered		(cF)

*Rhombohedral crystals are also described using hexagonal axes.

Most metals of engineering importance crystallize in either cubic or hexagonal forms, in one of three types of space lattice.

1. **Body-Centered Cubic (bcc).**
   Chromium, tungsten, alpha (α) iron, delta (δ) iron, molybdenum, vanadium and sodium exhibit this kind of lattice.
2. **Face-Centered Cubic (fcc).**
   This structure is more densely packed than bcc. Aluminum, nickel, copper, gold, silver, lead, platinum and gamma (γ) iron exhibit this kind of lattice.
3. **Hexagonal Close-Packed (hcp).**
   Magnesium, beryllium, zinc, cadmium and hafnium exhibit this kind of lattice.

SOLIDIFICATION

Associated with alloying is the subject of solidification; we shall briefly discuss this in the following paragraphs.

Lever Rule of Solidification

The calculation of the equilibrium solidification according to the Lever rule is a one-dimensional stepping calculation in which one of the variables, the

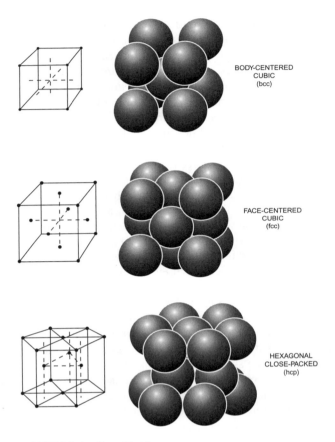

FIGURE 1-3-1   Crystal Lattices.

temperature, is stepped. At each step a single point equilibrium calculation is carried out. The results of the calculation are the phase compositions, phase fractions and enthalpies. Since complete diffusion is assumed for all phases, the calculation may be continued after the liquid phase has disappeared to determine changes in the solid phases.

The details of phase diagrams are discussed in Chapter 4 of this book. However it is important to clarify a few things here related to the use of Lever law and to proceed with the discussion of solidification and constitutional cooling.

We will also consider the utility and importance of phase diagrams, including the determination of the amount or proportion of each phase present, and the evolution of the microstructure of the sample during cooling.

In alloys there is unlimited solid solubility in both the liquid and solid phase and the process of freezing (or melting) does not occur at a single fixed temperature, as happens with pure metallic elements. Alloys freeze or melt over a range of temperatures. At each temperature between the liquidus and solidus lines, the equilibrium structure is a mixture of the two (solid and liquid) phases, with fixed amounts and compositions of each phase. These values can be determined from the phase diagram.

There are only two possibilities in any binary (two-component) alloy system. A point on the phase diagram may lie within a single-phase field, in which there are two degrees of freedom, and the temperature and composition can be varied with no change in the microstructure. Single-phase fields are usually marked on phase diagrams with either a name or a Greek letter. This also explains why tie lines are not used in a single-phase alloy system. Alternatively, they may lie within a two-phase field, in which the proportion of the two phases will vary with composition or temperature. It is in these regions that the tie line is used and the Lever law is applied.

Thus the phase diagrams can be used to determine what phase or phases are present at a particular overall composition and temperature. This will be limited by the fact that in a single phase region, there is just one phase and in a two-phase field, the ends of the tie line identify the presence of a second phase. The diagram also helps us to understand what the compositions of the phases will be. In a single-phase system the diagram gives the bulk composition, while the two-phase system is defined by the tie line.

The diagram also confirms that the single-phase system has 100% of the composition of one metal and it directs us to use the Lever law equation to determine the proportional composition of metals in a two-phase system.

The diagram also tells us the arrangements of these phases into microstructures as they are developed in equilibrium (slow) cooling.

Thus, by using Lever's equation, we can determine at any given temperature the phase(s) that are present and their compositions. We can even slice the cooling process and determine the phase(s) and composition at any given point in the temperature range.

CONSTITUTIONAL SUPERCOOLING

Solute is partitioned into liquid of solidification front. This causes a corresponding variation in the liquidus temperature, i.e., the temperature below which freezing begins. There is a positive temperature gradient in the liquid, giving rise to a supercooled zone of liquid ahead of the interface. This is called constitutional supercooling because it is caused by changes in composition.

A small perturbation on the interface will therefore expand into supercooled liquid. This gives rise to dendrites.

It follows that a supercooled zone only occurs when the liquidus temperature gradient at the interface is larger than the temperature gradient.

In practice it is very difficult to avoid constitutional supercooling, because the required velocity is very small. Directional solidification with a planar front is possible only at low growth rates. A good example of this is the production of a single crystal of silicon. In most cases the interface is unstable.

Mixing in liquids occurs by convection, reducing solute gradients in the liquid. The uniform composition throughout the solidification in the liquid is the exception to the rule.

A key phenomenon in solidification is the transfer of heat, which occurs by radiation, direct contact with the mold, conduction through air, and convection as referred to above in the air gap between the mold and ingot. A casting situation may be classified by whether or not significant thermal gradients are set up in the solidifying metal.

When a liquid metal's temperature is dropped sufficiently below its freezing point, stable aggregates of atoms or nuclei appear spontaneously at various points in the liquid. These solid nuclei act as centers for further crystallization.

As cooling continues, more atoms attach themselves to already existing nuclei or form new nuclei. Crystal growth continues in three dimensions with the atoms attaching themselves in certain preferred directions along the axes of the crystal. This forms the characteristic tree-like structure called a dendrite. Each dendrite grows in a random direction until finally the arms of the dendrites are filled and further growth is obstructed by the neighboring dendrite. As a result, the crystals solidify into irregular shapes and are called grains.

The mismatched area along which crystals meet is called the grain boundary. It has a non-crystalline or amorphous structure with irregularly spaced atoms. Because of this irregularity, grain boundaries tend to be regions of higher energy, and reactions such as corrosion and crack propagation are often associated with grain boundary sites. Nucleation is a major factor in determining the material's final properties.

ELEMENTARY THEORY OF NUCLEATION

Phase fluctuations occur as random events due to the thermal vibration of atoms. An individual fluctuation may or may not be associated with a reduction in free energy, but it can only survive and grow if there is such a reduction. There is a cost associated with the creation of a new phase, the interface energy, a penalty which becomes smaller as the particle surface to volume ratio decreases. In a metastable system this leads to a critical size of fluctuation beyond which growth is favored.

Let's consider a homogeneous nucleation of α from γ, for a spherical particle of radius r with an isotropic interfacial energy $\sigma_{\alpha\gamma}$, the change in free energy (ΔG) as a function of radius is:

$$\Delta G = 1.33\pi r^3 \Delta G_{CHEM} + 1.33\pi r^3 \Delta G_{STRAIN} + 4\pi r^2 \sigma_{\alpha\gamma} \qquad (1)$$

where:

$\Delta G_{CHEM} = G^{\alpha}v - G^{\gamma}v$, G_V is the Gibbs free energy per unit volume of α and

G_{STRAIN} is the strain energy per unit volume of α.

The maximum variation in the curve of ΔG is determined by differentiating equation (1) with respect to the radius r:

$$\Delta(\Delta G)/\Delta r = 4\pi r^2 [\Delta G_{CHEM} + \Delta G_{STRAIN}] + 8\pi r\sigma_{\alpha\gamma} \qquad (2)$$

In this equation the value of r* is obtained by setting the value to zero:

$$r^* = 2\sigma_{\alpha\gamma}/\Delta G_{CHEM} + \Delta G_{STRAIN} \qquad (3)$$

By substitution the value of r* into equation (1) above, the activation energy is obtained as:

$$G^* = 16\pi\sigma_{\alpha\gamma}^3/3\,(\Delta G_{CHEM} + \Delta G_{STRAIN})^2 \qquad (4)$$

The key outcome is that in classical nucleation the activation energy varies inversely with the square of the driving force, and the mechanism involves random phase fluctuations. The nucleation rate per unit volume, I_V will depend on the attempt frequency v, the number density of nucleation sites N_V and the probability of successful attempts. There is also a barrier Q to the transfer of atoms across the interface:

$$I_v = N_v{}^v \exp\{-G^*/kT\}\exp\{Q/kT\} \qquad (5)$$

ALLOTROPY

Some metals and alloys exhibit different crystalline lattices at different temperatures – a phenomenon called allotropy. Allotropy is a very important property of materials; these allotropic changes are the basis for heat treatment of many engineering materials.

Iron and its alloys are the most common engineering materials that have allotropic forms we will use this material as the basis of our discussion. Pure iron in its solid state has three allotropic forms, called austenite (y), ferrite and ϵ-iron. The latter has a hexagonal close-packed crystal structure, is the highest density state of iron, and is only stable at very high pressures. At ambient pressures, ferrite is stable at temperatures just below the equilibrium melting temperature (in which case it is called δ-iron) and at relatively low temperatures as the α form of iron. Austenite is the stable form in the intervening temperature range between δ and α. As was recognized a long time ago by Zener and others, this complicated (but useful) behavior is related to electronic and magnetic changes within the material as a function of temperature.

The phase behavior of pure iron does not change radically with the addition of small amounts of solute, i.e., in low-alloy steels. Using the example of a lightly alloyed steel weld, deposits begin to solidify with the epitaxial growth of delta-ferrite (δ) from the hot grain structure of the parent plate at the fusion boundary. The large temperature gradients at the solid/liquid interface ensure that solidification proceeds along a cellular front so that the final δ grains are columnar in shape, with the major axes of the grains lying roughly along the direction of maximum heat flow. On further cooling, austenite allotriomorphs nucleate at the δ-ferrite grain boundaries, and their higher rate of growth along the $\delta - \delta$ boundaries and, presumably, along temperature gradients, leads to the formation of columnar austenite grains whose shape resembles that of the original solidification structure.

It is important to state here that cooling metallurgy is an important part of welding, since welding involves a moving heat source in which the orientation of the temperature isotherms alter with time. Consequently, the major growth direction of the austenite is found to be somewhat different from that of the δ grains.

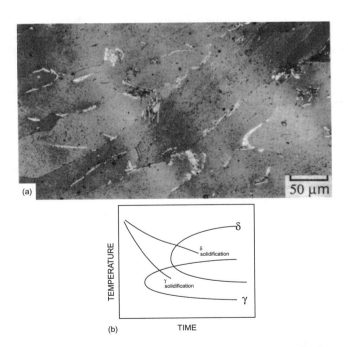

FIGURE 1-3-2   (a) Columnar δ ferrite grains, with austenite (light phase) allotriomorphs growing at the δ grain boundaries. (b) Schematic continuous cooling transformation diagram illustrating the solidification mode in low-alloy steels, as a function of the cooling rate. Faster cooling rates can in principle lead to solidification to metastable austenite.

If the cooling rate is high enough, then the liquid can be induced to solidify as metastable austenite as shown in Figure 1-3-2b. This could happen even when δ-ferrite is the thermodynamically favored phase in low-alloy steels. It has been suggested that this is especially likely when the partition coefficient $k = cS/cL$ is closer to unity for austenite than for ferrite. In this expression, cS and cL are the solute solubility in the solid and liquid phases respectively. In those circumstances the austenite growth rate can exceed that of δ-ferrite if the liquid is sufficiently undercooled. Solidification with austenite as the primary phase becomes more feasible as the steel is alloyed with austenite-stabilizing elements, until this eventually becomes the thermodynamically stable phase.

Solidification to austenite can be undesirable for two reasons; large inclusions tend to become trapped preferentially at the cusps in the advancing solid/liquid interface and end up at the columnar grain boundaries. When austenite forms directly from the liquid, the inclusions are located in the part of the weld that in the final microstructure corresponds to relatively brittle allotriomorphic ferrite. This is not the case with δ solidification, since during subsequent transformation, the daughter austenite grains cut across the δ / δ grain boundaries, leaving the large inclusions inside the grains where they can do less harm, and perhaps also be of use in stimulating the nucleation of acicular ferrite. The second reason to avoid solidification diffusion rates of substitutional elements is that orders of magnitude are larger in ferrite than in austenite. When this happens any segregation is less likely to persist when the liquid transforms to ferrite.

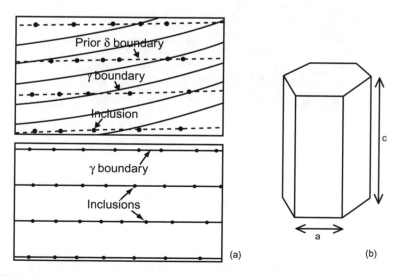

(a)   (b)

FIGURE 1-3-3   (a) Location of large inclusions for solidification as δ ferrite and as austenite. (b) A hexagonal prism model for the columnar austenite grains typical of the microstructure of steel weld metals.

CRYSTAL IMPERFECTIONS

Crystals are not always perfect; imperfections exist on the atomic scale, and are called vacancies. Vacancies are empty atom sites. The lattice planes around a vacancy are distorted. The converse of vacancies are the interstitials, which are extra atoms in the spaces of the lattice structure. Interstitials also produce lattice distortion.

Dislocations are disturbed regions between two substantially perfect parts of a crystal. Dislocations affect the tensile and compressive properties of metal.

GRAIN SIZE

Most engineering materials are polycrystalline, meaning that grain boundaries are an important feature of the microstructure, which can be manipulated to control the mechanical properties. For example the strength σ increases as the gain size d is reduced. Hall-Petch demonstrated this in his equation:

$$\sigma = \sigma_o + kd^{-1/2} \quad \text{(Hall-Petch equation)}$$

In the above relationship, σ_o and k are functions of the atomic micro-structure. Grain size refinement is the only mechanism that simultaneously improves strength and toughness, giving the metal its ability to absorb energy during fracture.

We have learned that the grain boundaries are defects which give an easy diffusion path. This implies that at high temperatures they would weaken the material, by allowing the easy diffusion of atoms in a way that leads to permanent creep. For elevated temperature applications it is necessary to minimize the amount of grain boundary area per unit volume. This is the very reason that the turbine blades for jet engines are made of single crystals, eliminating the formation of grain boundaries, and reducing the possibility of a weak phase.

The grain size can be controlled during the solidification stage by use of inoculants. A widely used alternative method is recrystallization.

The relationship between the rate of growth and the rate of nucleation determines the size of grains in a casting. The cooling rate is the most important factor in determining grain size. Rapid cooling allows many nuclei to be formed, resulting in a fine-grained material.

Insoluble impurities promote nucleation and promote fine grains. Disturbance of the melt during solidification tends to break up crystals before they become very large. In general, fine-grained materials exhibit better toughness and are harder and stronger than coarse-grained materials. Relatively, coarse-grained material may be preferred for high temperature service, since there are fewer grain boundaries to provide high temperature activation sites for deleterious reactions such as oxidation.

Structure of Materials

The importance of grain size, and the existence of different phases of a material was introduced in the previous chapter, in which we noted that the properties of a material depend to a large extent on the type, number, amount, and form of the phases present, and altering these quantities can change these properties. In this chapter we will discuss phase diagrams, which explain the conditions under which each phase exists and the conditions under which phase changes can occur. The term 'phase' refers to the region of space that is occupied by a physically homogeneous material.

Another term that is associated with a physically homogenous material is 'equilibrium'. This is defined as the dynamic condition of physical, chemical, mechanical or atomic balance that appears to be in a condition of rest as compared with change.

Three variables are used to describe the state of a system in equilibrium:

1. Temperature
2. Pressure
3. Composition.

A phase diagram is a graphical representation of the temperature and composition limits of phase fields in an alloy. It may be in equilibrium, or in a metastable condition.

There are three types of equilibria; stable, metastable and unstable. Stable equilibria exist when the material is in its minimum energy condition. Metastable equilibria are possible only when external energy can be introduced to stabilize the material. The unstable state of equilibria is the state in which external energy is required to bring the material to either a metastable or stable state.

Applied Welding Engineering: Processes, Codes and Standards.

Phase diagrams assume a constant (ambient) pressure. Since equilibrium conditions do not normally exist during heating and cooling, phase changes tend to actually occur at temperatures slightly higher or lower than the phase diagram would indicate. Rapid variations in temperature can prevent normally occurring phase changes.

PHASE DIAGRAMS

Phase diagrams, also called equilibrium or constitutional diagrams, are usually plotted with temperature on the ordinate axis, and alloy composition as a weight percentage on the abscissa. Phase diagrams are useful for the efficient and cost-effective understanding of alloys in various stages of research, development and production. For example they are useful in various aspects of metallurgical studies, such as the development of new alloys, and their fabrication, heat treatment procedures for developing specific properties, and resolution of issues arising from application of a specific alloy.

The term 'phase' was first used by J. William Gibbs who laid down the following rule, called the 'phase rule', to describe the relationship between temperature, pressure and number of homogeneous regions (phases) and the number of components. The rule can be described as follows:

$$f = c - p + 2 \qquad (1)$$

where:

f is the number of independent variables – called degrees of freedom,
c is the number of components, and
p is the number of stable phases in the system.

DIFFERENT TYPES OF PHASE DIAGRAMS

Phase diagrams can depict changes at various temperatures, and or pressure, and they can involve one or several components, as depicted in Figure 1-4-1. They are annotated with Latin letters.

A unary diagram is of one component. It is a simple, two-dimensional depiction of phase changes that occur as the temperature and or pressure changes. The phase rule states that all three phases (solid, liquid or gas) can exist in stable equilibrium only at a single point on a unary diagram ($f = 1 - 3 + 2 = 0$). In this diagram the three phases co-exist and the equilibrium occurs at the interface between the two adjoining phases. The invariant equilibrium among all three phases occurs at a single point, called the triple point, where the three phases intersect. This point is called invariant because here all the external variables are fixed, in other words there are no degrees of freedom ($f = 0$). A change in either temperature or pressure can change the equilibrium and one or two of the phases may disappear.

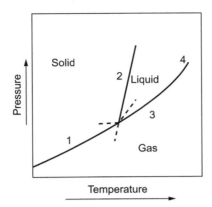

FIGURE 1-4-1   Pressure and Temperature Phase Diagram. *Reprinted with permission of ASM International. All rights reserved ASM.*

In contrast to the Invariant equilibrium discussed above are the Uninvariant and Bivariant equilibria.

A uninvariant or monovariant equilibrium refers two phases in a unary system that allows one degree of freedom ($f = 1 - 2 + 2 = 1$).

A bivariant equilibrium allows for either the temperature and pressure to be arbitrarily selected. This allows for the two degrees of freedom.

A binary diagram is of two components. In this system diagram, a third dimension is also added to the graph. Since most metallurgical problems are studied at one atmosphere of pressure, the graph is reduced to a two-dimensional plot of the temperature and composition. Since the pressure is held constant the phase rule changes.

$$f = c - p + 1 \qquad (2)$$

For a binary system with two components the stable equilibria could be one of the following:

- With one phase the degree of separation will be two.
- With two phases the degree of separation will be one.
- With three phases the degree of separation will be zero.

If the two components in a binary system have same crystal structure, and usually for two metals this is the case, and also similar atomic radii, then they are completely soluble in each other. These are also called miscible solids. The only type of solid phase formed will be a substitutional solid solution. In such a diagram, the boundary between the liquid field and the two-phase field is called liquidus and the space between the two-phase fields and solid field is called solidus. A series of cooling curves for various compositions is obtained by experiment and they are combined to form the phase diagram.

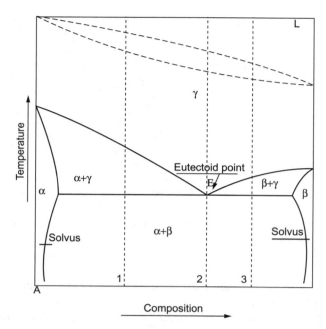

FIGURE 1-4-2 Iron Carbon Phase diagram. *Reprinted with permission of ASM International. All rights reserved.*

The upper line connecting the points showing the beginning of solidification is the liquidus line. The area above the liquidus line is a single-phase region of a homogeneous liquid solution. The lower line connecting the points showing the end of solidification is the solidus line. The area below is a single-phase region of a homogeneous solid solution.

The liquidus is the locus in a phase diagram that represents the temperature at which the alloy system begins to freeze on cooling or finish melting on heating. The solidus is the locus in a phase diagram that represents the temperature at which the alloy finishes freezing on cooling, or begins to melt on heating. When the phase is in the equilibrium field it is called the conjugated phase. When the liquidus and solidus meet tangentially at some point, a maximum and minimum are produced in the phase fields.

Ternary phase diagrams, as the name suggests, are diagrams that address the alloys consisting of three components. The phase diagram of such an alloy is plotted within a triangle. Figure 1-4-3 is a ternary phase diagram.

Each side of the triangle functions as the composition abscissa for a binary system, AB, BC, and CA. The composition for the given alloy at point X on the diagram is 20% A, 40% B and 40% C. To indicate the role of temperature

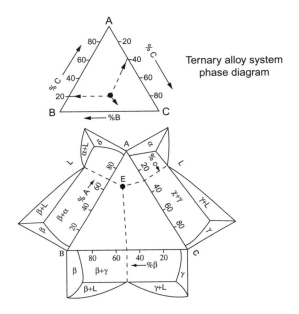

FIGURE 1-4-3 Ternary Alloy Phase Diagram.

and phase composition at a given temperature on a ternary diagram, it is necessary to move into the third dimension. This can be done in several different ways; diagrammatically a binary phase diagram can be drawn from each of the abscissa on the sides of the triangle, as shown in the figure for a eutectic alloy system. A ternary system does not facilitate the designation of temperature limits for phase fields. It is however possible to construct several binary phase diagrams to indicate at the fixed points the contents of one or the other of the components.

Other possible phase diagrams are listed below, but they will not be discussed in this book.

- A quaternary diagram is of four components.
- A quinary diagram is of five components.
- A sexinary diagram is of six components.
- A septenary diagram is of seven components.
- An octanary diagram is of eight components.
- A nonary diagram is of nine components.
- A decinary diagram is of ten components.

IRON-IRON CARBIDE PHASE DIAGRAM

Our discussion of phase diagrams would be incomplete without discussing the most often used and referenced phase diagram for industrial applications – the iron-iron carbide phase diagram.

Alloying iron with carbon produces a great variety of changes in the final material. These possibilities are commercially exploited by developing new and varied grades of steels with equally varied properties. Carbon is an element with an atomic weight of 12, which exists in three allotropic forms; firstly amorphous carbon in the form of lampblack and coal etc., secondly graphite, and finally diamond. The melting point of carbon varies but it is somewhere close to 3,300°C (6,000°F). This clearly indicates that carbon is present in iron in the solid state, and forms solutions in molten iron. The state of carbon in solid iron is very complex and varied. Carbon is an austenite former, because it extends the temperature range over which the face centered cubic (fcc) crystalline structure is stable. A significant amount of carbon is dissolved in gamma (γ) iron (fcc) and forms an intermetallic solution. The solubility of carbon in alpha (α) iron (bcc) is very limited and a maximum of 0.0218% can be retained at 727°C (1,340°F) in interstitial solution, but its solubility at room temperature drops to 0.008%. The carbon that cannot remain in solid solution in alpha iron or ferrite will form a compound, rather than exist as free carbon; this compound is iron carbide (Fe_3C), also called 'cementite'. This is the reason that the phase diagram of iron and carbon is called the iron-iron carbide diagram.

Carbon can exist in iron alloys as free carbon in the form of graphite at very high carbon content, such as gray cast iron. This is discussed in subsequent chapters.

EXPLANATION OF THE IRON-CARBON PHASE DIAGRAM

In Figure 1-4-4, the composition on the abscissa (horizontal direction) and temperature on ordinate (vertical direction) are correlated at atmospheric pressure. The weight percentage of components on the abscissa is plotted on a logarithmic scale, to allow the areas of interest to be expanded – that is the lower range of percentages of carbon – and the area of cast iron, which is of relatively low interest, to be compressed. The atomic percentage is indicated on the top horizontal line. Both Celsius and Fahrenheit temperature scales are used for proper understanding of the phase diagram.

The iron-iron carbide alloy has a eutectic and a peritectic reaction; the addition of carbon lowers the freezing point of alloys containing up to about 5% carbon, and causes solidification over a range of temperatures, which as discussed earlier is a typical effect of alloying.

As carbon is added, the liquidus steadily decreases along the curve ABC until a eutectic composition is reached at about 4.3% carbon. A further increase in carbon content causes the liquidus to rise along the curve CD. In general, the

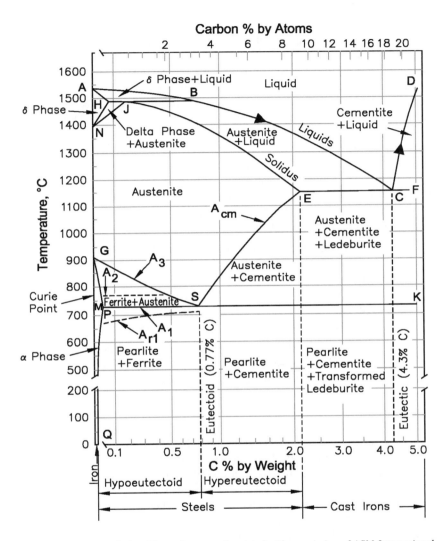

FIGURE 1-4-4   Iron Carbon Phase diagram. *Reprinted with permission of ASM International. All rights reserved.*

liquidus of steels is lowered over a relatively small temperature range, as the carbon content is increased up to the maximum that is incorporated in high-carbon steels. Cast irons however have a substantially lower liquidus compared to steels. This difference is clearly apparent when welding the two materials.

Most steels that are subject to welding usually contain up to 0.5% carbon, so there is reason to pay closer attention to this section of the phase diagram. In a steel-melt containing less than 0.53% carbon, the first crystals of solidification to appear below the liquidus AB are the δ (bcc) phase. The composition indicated by the line AH is part of the solidus AHJEF curve. The terms 'delta iron' and 'delta ferrite' are used to designate either iron that is free of carbon, or a solid solution of iron and carbon that has the bcc crystalline structure just below its solidus. At 1,495°C (2,723°F) alloys containing 0.09% to 0.53% carbon encounter the peritectic boundary. At over 0.53% carbon, the crystals that form as the temperature reaches the liquidus BC are called austenite. This phase is an interstitial solid solution of carbon in γ iron (fcc).

The line JE shows the composition of this primary austenite crystal. Austenite formed as the δ phase, along the line HN, during cooling is called secondary austenite. A summary of the important phases that are formed during solidification are listed below.

H   Delta (δ) phase (bcc) containing 0.09% carbon.
J   Gamma (γ) phase (fcc) containing 0.17% carbon. (Peritectic composition).
B   Peritectic liquid, containing 0.53% carbon.
HJB Peritectic temperature 1,495°C (2,723°F).
NJE Austenite (γ) phase (fcc) with carbon in interstitial solid solution.

Since the outcome of the reaction is not shown in the final microstructure of carbon and low alloy steels, little attention is paid to the peritectic reaction in solidification stages. In high alloys, however, the peritectic reaction becomes important because the reaction often does not proceed to completion, and small amounts of δ phase (bcc) that is of a different structure may be retained in a matrix.

During cooling below the liquidus line BC, the iron carbon composition that contains 2% and up to 4.3% carbon is first to form austenite. The carbon content of the austenite can be ascertained by applying the Lever law to the solidus JE. This primary solid will be lower in carbon than the overall alloy composition, and when the temperature reaches the eutectic line EC, the remaining liquid will be enriched to the eutectic carbon content of 4.3%. Final solidification at 1,154°C (2,109°F) the eutectic line on the diagram proceeds by formation of a eutectic consisting of metastable austenite and cementite. This structure is also called ledeburite. On continued cooling below the eutectic boundary PSK at 727°C (1,341°F) the metastable austenite undergoes a eutectoid reaction and decomposes into ferrite and cementite, a structure known as pearlite.

In compositions containing over 4.3% carbon, the first crystals appear when the temperature reaches the liquidus line, CD. During cooling crystals of cementite appear in the iron carbon compound Fe_3C, which contains 6.67% carbon. The following is a brief summary of the diagram.

ECF Part of the eutectic line, which extends to 6.67% carbon (outer limit 6.67% carbon is outside the diagram).

E   Eutectic austenite, 2.1% carbon.
C   Eutectic liquid, 4.3% carbon.

Along the line ECF, the three components austenite, eutectic liquid (E), and cementite (C) are in equilibrium. If heat is applied, austenite E and cementite react to form a liquid containing 4.3% carbon at a constant temperature of 1,145°C (2,109°F). If heat is removed, the eutectic liquid C discharges austenite E and cementite in a finally divided eutectic form. Liquid is not found below the solidus line AHJEF.

Besides the less significant lines NJ and NH, which encompass delta phase austenite, there are only five lines in the phase diagram that are below the solidus. We shall now discuss the two remaining points P and S.

Lines GS and SE are analogous to lines AC and CD, and the molten alloy-liquid solution is analogous to austenite-solid solution. When a solid solution of austenite reaches GS (Ar_3) on cooling, crystals of alpha iron appear, and the composition is known as ferrite. Without cementite and austenite support, ferrite can exist only in the area to the left of GPQ. When austenite reaches SE (A_{cm}) on cooling, crystals of cementite start to appear. Pertinent details of the eutectoid portion of the phase diagram are given below.

PSK Part of the eutectoid line 727°C (1,341°F).
P   Eutectoid ferrite, containing 0.0218% carbon.
S   The eutectoid, containing 0.77% carbon.

The right-hand end of the eutectoid line extends to 6.67% carbon, corresponding to the composition of cementite, Fe_3C. Along the line PSK, the material P, S, and cementite are in equilibrium. If heat is applied, P and cementite react to form austenite of composition S, containing 0.77% carbon, at a temperature of 727°C (1,341°F). If however the heat is removed, then S decomposes into ferrite (P) and cementite. The crystals of ferrite and cementite arrange themselves in a lamellar structure with a thin plate of cementite separating every pair of ferrite plates. This structure is called pearlite.

The transformation temperatures, identified as A_1 and A_3 for iron carbon alloys, can vary from those shown by the boundary drawn in the graph. If either the heating or cooling rate is faster than the very slow rate required to establish the equilibrium conditions, the transformation will occur at somewhat different temperatures.

RATIONALE FOR LETTER DESIGNATIONS IN THE IRON-IRON CARBIDE PHASE DIAGRAM

Two of the alphabetical designations in Figure 1-4-4 deserve further explanation:

- The M designation, which marks the change in magnetic properties of iron at 770°C (1 418°F), is the Curie point
- The A_2 designation indicates this same change in steel.

The importance of critical temperatures in dealing with steels has fostered the practice of marking not only the occurrence of these changes in the solid state, but also the sequence and conditions under which the specific temperature point was determined. If these critical temperatures are determined by plotting a curve, the letter A is employed. Etymologically the letter A comes from the French word 'arrêt' meaning stop. The transformation in the solid state can involve diffusion, and therefore require more time than in the liquid state. Hence it is important to indicate the conditions under which the temperature was recorded.

When a phase change is determined during a slow rate of heating, the critical temperature would be labeled A_c (the c referring to the French word 'chauffage' meaning heating). If the determination was made during slow cooling, the designation would be A_r (the letter r coming from another French word 'refroidissement' meaning cooling). When the critical temperature for phase transformation is established under equilibrium conditions, the designation is A_e (the letter e is for equilibrium).

The sequence of phase transformations in steel is viewed from their occurrence during heating, therefore the lowest critical point is marked as A_{C1}, and the subsequent points are A_{C2} and A_{C3}. The notation A_{cm} is used to designate the curve, which indicates the start of austenite transformation in hypereutectoid steel that has a carbon percentage >0.77% during cooling where cementite forms.

Production of Steel

Since we have discussed basic physical metallurgy, the behavior of iron and steel on heating and cooling, and the process of solidification and phase formation, it is now an appropriate time to study the steel-making process.

Steel making is an ancient process, and numerous developments have taken place in the technology over the many years of its use.

The two processes that are now commonly used for the production of steel are the Basic Oxygen Furnace (BOF) and the Electric Arc Furnace (EAF); they use variety of charge materials and technologies.

The Electric Arc Furnace (EAF) process uses virtually 100% old steel to make new steel. EAFs make up approximately 60% of today's steelmaking in the US The operational details of EAF and BOF processes are discussed below.

THE ELECTRIC ARC FURNACE (EAF) PROCESS

As the name implies, the process uses an electric arc furnace to melt the charge to make steel. It is a batch melting process producing batches of molten steel known as 'heats'. The electric arc furnace operating cycle is called the tap-to-tap cycle and is made up of the following operations:

1. Furnace charging
2. Melting
3. Refining
4. De-Slagging
5. Tapping
6. Furnace turn around

The tap-to-tap cycle time is generally less than 60 minutes; some double shell furnace operations are able to reduce the tap-to-tap time to about 30 to 40 minutes.

Furnace Charging

The first step in the production of any heat is to select the grade of steel to be made. Usually a schedule is developed prior to each production shift. Thus the melter will know the schedule for his shift in advance. The scrap yard operator will prepare buckets of scrap according to the needs of the melter. Preparation of the charge bucket is an important operation, not only to ensure proper melt-in chemistry but also to ensure good melting conditions. The scrap must be layered in the bucket according to size and density to promote the rapid formation of a liquid pool of steel in the hearth while providing protection for the sidewalls and roof from electric arc radiation. Other considerations include:

- Minimization of scrap cave-ins which can break electrodes
- Ensuring that large heavy pieces of scrap do not lie directly in front of burner ports which would result in blowback of the flame onto the water-cooled panels.

Generally a charge includes lime and carbon, but these can be injected later during the furnace operation. Often mills practice a combination of both.

The first step in any tap-to-tap cycle is 'charging' in the scrap. The roof and electrodes of the furnace are raised and swung aside, to allow the scrap charging crane to place a fully loaded scrap bucket to sit over the top of the furnace opening. The bucket's bottom operates like a clamshell, with two segments on the bottom retracting to let the scrap fall into the furnace. The roof and electrodes then swing back into place over the furnace. The roof and electrodes are lowered to strike an arc on the scrap. This commences the melting portion of the cycle. The number of charge buckets of scrap required to produce a heat of steel primarily depends on the capacity volume of the furnace and the scrap density.

Modern furnaces are designed to operate with a minimum of back-charges. This is advantageous because charging time is dead-time for the furnace. Reducing dead-times allows the production time of the furnace to increase, and reduces energy loss. Dead-time in a furnace can amount to about

10–20 kWh/ton for each occurrence. Generally about 2 to 3 buckets of scrap is charged for each heat.

Melting

The EAF has developed as an efficient melting apparatus, with designs focusing on increased capacity. Melting is accomplished by supplying energy to the furnace interior. This energy can be electrical or chemical. Electrical energy is supplied via graphite electrodes, and is usually the largest contributor to melting operations. Initially, an intermediate voltage tap is selected until the electrodes dig into the scrap. Usually, light scrap is placed on top of the charge to accelerate bore-in. Nearly 15% of the scrap is melted during this initial bore-in period. After a few minutes, the electrodes penetrate the scrap deep enough to allow for a high voltage tap that can develop a long arc without fear of radiation damage to the roof. The long arc maximizes the heat of the arc, as it transfers the power to the scrap and a liquid pool of metal, in the hearth of the furnace.

If the initiation of the arc is erratic and unstable, wide swings in current are observed accompanied by rapid movement of the electrodes. As the temperature in the furnace rises the arc stabilizes. As the molten pool is formed, the arc becomes quite stable and the average power input increases.

Chemical energy is supplied via several sources, including oxy-fuel burners and oxygen lances. Oxy-fuel burners burn natural gas using oxygen or a blend of oxygen and air. Heat is transferred to the scrap by flame radiation and convection by the hot products of combustion. Heat is transferred within the scrap by conduction. Large pieces of scrap take longer to melt into the bath than smaller pieces. In some operations, oxygen is injected via a consumable pipe lance to 'cut' the scrap. The oxygen reacts with the hot scrap and burns iron to produce intense heat for cutting the scrap. Once a molten pool of steel is generated, oxygen can be lanced directly into the molten bath of steel. The oxygen reacts with several components in the bath, including aluminum, silicon, manganese, phosphorus, carbon and iron. All of these reactions are exothermic and supply additional energy to aid in the melting of the scrap. The metallic oxides that are formed end up in the slag. The reaction of oxygen with carbon in the bath produces carbon monoxide, which either burns in the furnace if there is sufficient oxygen, and/or is exhausted through the direct evacuation system where it is burned and conveyed to the pollution control system.

Once enough scrap has been melted to accommodate the second charge, the charging process is repeated. Once the final scrap charge is melted, the furnace sidewalls are exposed to intense radiation from the arc. As a result, the voltage is reduced. Alternatively, creation of a foamy slag allows the arc to be buried, which protects the furnace shell. In addition, a greater amount of energy is retained in the slag and is transferred to the bath resulting in greater energy efficiency.

Once the final scrap charge is fully melted, flat bath conditions are reached. At this point, the bath temperature is measured and a sample is collected. The analysis of the bath chemistry allows the melter to determine the amount of oxygen to be blown during refining. At this point, the melter can also start to arrange for the bulk tap alloy additions to be made. These quantities are finalized after the refining period.

Refining

Traditionally, refining operations in the electric arc furnace involve the removal of phosphorus, sulfur, aluminum, silicon, manganese and carbon from the steel. However new developments in the understanding of the role of dissolved gases, especially hydrogen and nitrogen, make it essential that these gases are also reduced to acceptable limits. These gases are not removed during the refining operations once a flat bath is achieved. The presence of oxygen during flat bath operation reduces the carbon content to the desired level for tapping. Most of the compounds that are to be removed during refining have a higher affinity for oxygen than the carbon. Thus the oxygen reacts preferentially with these elements to form oxides, as they float out of the steel and through the slag.

In modern EAF operations, especially those operating with a 'hot heel' of molten steel and slag retained from the prior heat, oxygen is often blown into the bath. This technique allows for the simultaneous operation of melting and refining in the furnace.

Phosphorus Removal

Excessive concentrations of phosphorus and sulfur are normally present in the furnace charge, which must be reduced to acceptable limits. Unfortunately the conditions that favor the removal of phosphorus are the opposite of those for the removal of sulfur. Therefore once these materials are pushed into the slag phase they can revert back into the steel. Phosphorus retention in the slag is a function of the following factors.

1. Bath temperature
2. The slag basicity and
3. FeO levels in the slag

At higher temperature or low FeO levels, the phosphorus reverts from the slag back into the bath. Phosphorus removal is carried out as early as possible in the heat. Hot heel practice is very beneficial in phosphorus removal because oxygen can be lanced into the bath while its temperature is quite low. Early in the heat the slag contains high levels of FeO as it is carried over from the previous heat; this helps removal of phosphorus. High slag basicity (i.e. high lime content) is also beneficial for phosphorus removal. This method has some limitation, as saturating the slag with lime increases its viscosity, making it less effective. To increase the fluidity of slag (fluidize the slag) fluorspar is added. Stirring the bath with inert gas is also beneficial, because it renews the slag to metal interface thus improving the reaction kinetics.

In general, if low phosphorus levels are needed for a particular steel grade, the scrap is selected to give a low level at melt-in; usually the phosphorus is reduced by 20% to 50% in the EAF process.

Sulfur Removal

Sulfur is removed mainly as a sulfide dissolved in the slag. The sulfur partition between the slag and metal is dependent on slag chemistry and is favored at low steel oxidation levels. Removal of sulfur in the EAF is difficult, especially in modern practices where the oxidation level of the bath is high. Generally the partition ratio is between 3 and 5 for EAF operations. Most operations find it more effective to desulphurize during the reduction phase of steelmaking. This means that desulphurization is performed during tapping (where a calcium aluminate slag is produced) and during ladle furnace operations. For reducing conditions where the bath has a much lower oxygen activity, distribution ratios for sulfur of between 20 and 100 can be achieved.

Control of the metallic constituents in the bath determines the properties of the final product. Usually, the melter aims at lower levels in the bath than are specified for the final product. Oxygen reacts with aluminum, silicon and manganese to form metallic oxides, which are slag components. These metallic oxides tend to react with oxygen before the carbon. They also react with FeO resulting in a recovery of iron units to the bath. For example:

$$Mn + FeO = MnO + Fe$$

Manganese is typically lowered to about 0.06% in the bath. The reaction of carbon with oxygen to produce carbon monoxide (CO) is an important step, as it supplies a less expensive form of energy to the bath, and performs several important refining reactions. In modern EAF operations, the combination of oxygen with carbon can supply between 30% and 40% of net heat input to the furnace. Evolution of carbon monoxide is very important for slag foaming. Coupled with a basic slag, CO bubbles are trapped in the slag, causing it to 'foam' and helping to bury the arc. This greatly improves thermal efficiency and allows the furnace to operate at high arc voltages even after a flat bath has been achieved. Burying the arc also helps to prevent nitrogen from being exposed to the arc where it can dissociate and enter into the steel.

Nitrogen and Hydrogen Control

If the CO gas is evolved within the steel bath, it helps to strip nitrogen and hydrogen from the steel. At 1,600°C, the maximum solubility of nitrogen in pure iron is 450 ppm. Typically, nitrogen levels in the steel are 80–100 ppm following tapping. Nitrogen levels as low as 50 ppm can be achieved in the furnace prior to tap. Bottom tapping is beneficial for maintaining low nitrogen levels because tapping is fast and a tight tap stream is maintained. High oxygen potential in the steel is beneficial for low nitrogen levels and the heat should be tapped open as opposed to blocking the heat.

Decarburization is also beneficial for the removal of hydrogen. It has been demonstrated that decarburizing at a rate of 1% per hour can lower hydrogen levels in the steel from 8 ppm down to 2 ppm in 10 minutes.

At the end of refining, a bath temperature measurement and a bath sample are taken. If the temperature is too low, power to the bath may be increased. This is not a big concern in modern mills, where temperature adjustment is carried out in the ladle furnace.

De-Slagging

De-slagging operations are carried out to remove impurities from the furnace. During melting and refining operations, some of the undesirable materials within the bath are oxidized and enter the slag phase.

It is advantageous to remove as much phosphorus into the slag as early in the heat as possible (i.e. while the bath temperature is still low). The furnace is then tilted backwards, and slag is poured out of the furnace through the slag door. Removal of the slag at this stage eliminates the possibility of phosphorus reversion.

Phosphorus reversion is a phenomenon that occurs during slag foaming operations. Carbon may be injected into the slag where it will reduce FeO to metallic iron and in the process produce carbon monoxide to help foam the slag. If the high phosphorus slag has not been removed prior to this operation, phosphorus reversion will occur. During slag foaming, slag may overflow the steel level in the EAF and flow out of the slag door.

The slag contains various oxides and also some elements that are harmful to the steel. A typical analysis of Electric Arc Furnace slag is given in Table 1-5-1.

Tapping

Once the desired steel composition and temperature are achieved in the furnace, the tap-hole is opened, the furnace is tilted, and the steel is poured into a ladle for transfer to the next operation. During the tapping process bulk alloy additions are done, based on the bath analysis and the desired steel grade. Prior to further processing, de-oxidizers are added to the steel to lower the oxygen content. This is commonly referred to as 'blocking the heat' or 'killing the steel'. Common de-oxidizers are aluminum or silicon in the form of ferrosilicon or silicomanganese. Most carbon steel operations aim for minimal slag carry-over. A new slag cover is 'built' during tapping. For ladle furnace operations, a calcium aluminate slag is a good choice for sulfur control. Slag

TABLE 1-5-1 Describing Typical Slag Composition of Electric Arc Furnace (EAF)

Component	Source	Composition Range
CaO	Charged	40–60%
SiO$_2$	Oxidation product	5–15%
FeO	Oxidation product	10–30%
MgO	Charged as dolomite	3–8%
CaF$_2$	Charged – slag fluidizer	
MnO	Oxidation product	2–5%
S	Absorbed from steel	
P	Oxidation product	

forming compounds are added to the ladle at tap so that a slag cover is formed prior to transfer to the ladle furnace. Additional slag materials may be added at the ladle furnace if the slag cover is insufficient.

BASIC OXYGEN FURNACE (BOF)

The Basic Oxygen Furnace process uses 25 to 35 percent old steel to make new steel. The oxygen steel-making process is one of the most common and efficient steel production methods. A brief discussion of the process is included in the following paragraphs.

The oxygen steel-making process is a generic name given to those processes in which gaseous oxygen is used as the primary agent for autothermic generation of heat as a result of the oxidation of dissolved impurities like carbon, silicon, manganese and phosphorus, and to a limited extent the oxidation of iron itself. Several types of oxygen steel-making processes are practiced, including top blowing, bottom blowing and combined blowing.

The essential features of conventional steel-making are the partial oxidation of the carbon, silicon, phosphorus and manganese present in pig iron and the accompanying reduction in sulfur levels. Blast furnace hot metal for the Linz-Donawitz (LD) Basic Oxygen Furnace steel-making process ideally has following analysis:

- Carbon = 4.2%
- Silicon = max 0.8%
- Manganese = max 0.8%
- Sulfur = max 0.05%
- Phosphorus = max 0.15%

These solute elements are diluted by the addition of scrap, which forms some 20–30% of the metallic charge.

Refining Reactions

In the basic LD oxygen steel-making process, the oxygen required for the refining reactions is supplied as a gas, and both metal and slag are initially oxidized according to the following reactions.

$$\tfrac{1}{2}O_{2(g)} \leftrightarrow [O] \tag{1}$$

$$Fe + [O] \leftrightarrow (FeO) \tag{2}$$

$$2(FeO) + \tfrac{1}{2}O_{2(g)} \leftrightarrow (Fe_2O_3) \tag{3}$$

Carbon

The actual distribution of oxygen between slag and metal is not easily determined since it is a function of a number of variables including lance height and oxygen flow rate. The principal refining reaction is of course the removal of carbon as described by following reactions:

$$[C] + [O] \leftrightarrow CO_2 \tag{4}$$

$$[C] + (FeO) \leftrightarrow CO_2 + Fe \tag{5}$$

Figure 1-5-1 is an idealized diagram showing the changes in concentrations of the elements in an LD metal bath during oxygen blowing. The basic thermodynamic data for these reactions are well established, and the equilibrium carbon and oxygen contents may be readily calculated for all temperatures and pressures encountered in steelmaking.

Oxidation of carbon during the oxygen converter process is most important, since the reaction increases the temperature and evolves a large amount of CO and CO_2 gases that cause agitation of metal and slag and helps remove hydrogen, nitrogen and part of the non-metallic inclusions from the metal. Owing to the pressure of the oxygen supplied and the evolution of large quantities of gases, the liquid bath becomes an intimate mixture of slag, metal and gas bubbles, with an enormous contact surface. Because of this, the oxidation of carbon self-accelerates and proceeds extremely rapidly.

Silicon

In accordance with thermodynamic predictions, the removal of silicon is usually completed relatively early in the blow. The typical reaction is represented by equations (6) and (7).

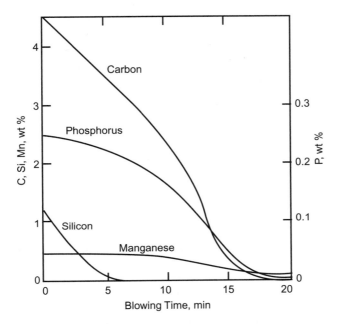

FIGURE 1-5-1   The changes of bath composition during the blow in a basic oxygen steelmaking converter (idealized).

$$[Si] + 2[O] \leftrightarrow (SiO_2) \tag{6}$$

$$[Si] + 2(FeO) \leftrightarrow (SiO_2) + 2Fe \tag{7}$$

Manganese

Similar equations can be applied to manganese removal:

$$[Mn] + [O] \leftrightarrow (MnO) \tag{8}$$

$$[Mn] + (FeO) \leftrightarrow (MnO) + Fe \tag{9}$$

Initially, the level of manganese in the bath falls due to oxidation, but later a slight reversion occurs, followed by a second fall in level. These changes in the manganese content of the bath are attributed to the combined effects of rising temperature and variable slag composition, and reactions of manganese and ferrous oxides, suggesting that the reaction is close equilibrium. This view is supported by the observation that at the end of blowing, the manganese content is found to be about 82% of the equilibrium value when lump lime is used and 85% of the equilibrium value when powdered lime is injected.

In the middle part of blow, the (FeO) level in the slag falls as a consequence of the decarburization process and the dilution that accompanies lime fluxing. However, towards the end of the blow, the (FeO) increases again, as carbon removal becomes less intense and dilution begins to affect the activity of manganese oxide, with the result that manganese transfers from bath to slag. To some extent raising the temperature may minimize manganese loss.

Phosphorus

The partitioning of phosphorus between slag and metal is known to be very sensitive to process conditions, and some research laboratories associated with steel mills have produced a kinetic model based on simple assumptions.

Healy has reviewed the distribution of phosphorus between slag and metal. He concluded that the thermodynamic behavior of phosphorus is best explained by a modified version of the ionic theory first proposed by Flood and Grjotheim. The slag-metal reaction is written in ionic form in the equation below:

$$2[P] + 5[O] + 3\,(O^{2-}) \leftrightarrow 2(PO^{3-})_4 \tag{10}$$

Healy expressed the equilibrium distribution of phosphorus by equations that apply to specific concentration ranges in the $CaO\text{-}SiO_2\text{-}FeO$ system:

$$\log\,(\%P)/[P] = 22\,350/T + 7\,\log\%CaO + 2.5\,\log Fe_t - 24.0 \tag{11}$$

$$\log\,(\%P)/[P] = 22\,350/T + 0.08\,\log\%CaO + 2.5\,\log Fe_t - 16.0 \tag{12}$$

Equation (11) applies to slag with $CaO > 24\%$ and equation (12) is for slag containing CaO from zero to saturation level.

In practice the phosphorus partition ratios are far from the values calculated for equilibrium with carbon-free iron, because the oxygen potential of the slag-metal system is influenced by decarburization. A limited correlation with the carbon content in the bath has been reported, although other mills have suggested that extensive dephosphorization should be possible at high carbon levels, provided that the slag is sufficiently basic.

On the other hand, the dependence of phosphorus distribution on the FeO content of slag in LD and Q-BOP is shown in Figure 1-5-2. A parameter k_{PS} is defined as:

$$k_{PS} = (\%P_2O_5)/[\%P] \cdot (1 + (\%SiO_2)) = \varphi((\%FeO), B) \tag{13}$$

Where:
B is basicity. For $B > 2.5$, k_{PS} is found to be independent of B.

The distribution of phosphorus is also found to be related to the content of carbon in steel at the time of tapping; owing to lower carbon levels achieved

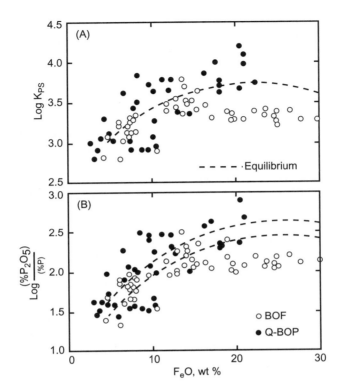

FIGURE 1-5-2   Effect of FeO content of slag on phosphorus distribution and log k_{PS} value.

in bottom-blown process, the phosphorus distribution is expected to be better than in LD. In general, high basicity and the low temperature of slag (irrespective of the FeO content) favor dephosphorization.

Sulfur Removal

Sulfur transfer takes place through the following reactions:

$$[S] + (O_2)_g = (SO_2)_g \tag{14}$$

It is found that approximately 15–25% of dissolved sulfur is directly oxidized into the gaseous phase due to the turbulent and oxidizing conditions existing in the jet impact zone.

In the Basic Oxygen Furnace, the metal desulfurization proceeds slowly because it is a diffusion process. It may be accelerated to an extent by improving the bath mixing and increasing the temperature, fluidity and basicity of the slag, and the activity of sulfur. At the initial stage of the heat, when the metal is rich in carbon and silicon, the activity of sulfur is high. Besides, part of sulfur is removed at the initial stages of the process when the temperature of melt is still relatively low, via its reaction with manganese, as the following reaction suggests:

$$[Mn] + [S] = (MnS) \tag{15}$$

A rise in the concentration of iron oxides in the slag promotes dissolution of lime, which favors desulfurization. The secondary and most intensive desulfurization occurs at the end, and at the heat when the lime dissolves in the slag with a maximum rate and the slag basicity (B) reaches 2.8 or more. Thus the total desulfurization of the metal is mainly decided by the basicity of the homogeneous final slag, which is formed in the oxygen converter process during the last minutes of metal blowing.

At higher slag basicity, the residual concentration of sulfur in metal bath becomes lower, allowing the coefficient of sulfur distribution between slag and metal to be raised up to 10. The greater the bulk of slag is the largest part of the sulfur will pass into slag at the same sulfur distribution coefficient. But there is limit to the benefits of a very large bulk of slag, since this increases the iron loss due to burning, causes splashing and rapid wear of the lining.

In a steel plant, regression equations based on operational data are employed to predict the end point for sulfur within acceptable limits. Equation (16) is an example:

$$(\%S)/[S] = 1.42B - 0.13(\%FeO) + 0.89 \tag{16}$$

The equation shows the beneficial influence of slag basicity and the retarding influence of FeO on sulfur distribution. A large number of such correlations are reported in the literature, but they are only locally applicable.

DEOXIDATION OF STEEL

Deoxidation is the process that allows the removal of excess oxygen from molten metal. The procedure involves adding carefully measured amounts of materials that have a high affinity for oxygen. They attract the oxygen present in the molten steel and form oxides, thus reducing the oxygen content of the steel. These oxides are either removed in gaseous form or they readily form slag, and float on the molten steel surface to be skimmed off. The deoxidation of steel is usually performed by adding manganese (Mn), silicon (Si) and aluminum (Al); other deoxidizers used are chromium (Cr), vanadium (V), titanium (Ti), zirconium (Zr) and boron (B). With the exception of chromium and special alloy-steels the presence of these elements is not desirable, hence they are very rarely used, and often in combination with the primary deoxidation elements given above.

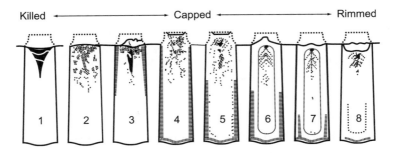

Killed ←————————————→ Capped ←————————————→ Rimmed

FIGURE 1-5-3   Series of typical ingot structures.

As previously stated, the deoxidizing agents have high a affinity with oxygen. This essential property of deoxidizing agents is exploited to remove excess oxygen from molten steel, but if not carefully controlled the same property presents a paradox, since if the concentration of deoxidizer in the melt increases over a critical limit, the process reverses to the reoxidation of steel.

Deoxidation is the last stage in the steel-making process. In the Basic Oxygen Furnace (BOF) and other similar steel-making practices, the steel bath at the time of tapping contains 400 to 800 ppm activity of oxygen. Deoxidation is carried out during tapping by adding appropriate amounts of ferromanganese, ferrosilicon and/or aluminum or other special deoxidizers into the tap-ladle. If at the end of the blow the carbon content of the steel is below specifications, the metal is also recarburized in the ladle. However, such last minute additions are for fine adjustment; large additions in the ladle are undesirable, because the process brings down the temperature of the steel bath. The steel is often poured into fixed molds, to be cast as ingots. In a continuous casting process, the molds are a form that is in continuous movement and the solidification takes place while the mold is in motion.

Eight typical conditions of commercial ingots, cast in identical bottle-top molds, in relation to the degree of suppression of gas evolution are shown schematically in Figure 1-5-3 above. The dotted line on the top indicates the height to which the steel originally was poured in each ingot mold. Depending on the carbon and oxygen content, the ingot structure ranges from a fully killed or dead-killed ingot (N°1 in Figure 1-5-3) to a violently rimmed ingot (N°8). Included in the series are killed steel (marked N°1), semi-killed steel (marked N°2), capped steel (N°5), and rimmed steel (N°7).

Rimmed Steel

Rimmed steels are usually tapped without additions of deoxidizers to the steel in the furnace or only small additions to the molten steel in the ladle. This is to ensure sufficient presence of oxygen to facilitate gas evolution by reacting in the mold with carbon. The exact procedures followed depend upon whether the steel has a carbon content in the ranges of 0.12–0.15%, or in the lower range of ≤0.10%. When the metal in the ingot mold begins to solidify, there is a brisk evolution of carbon monoxide, resulting in an outer ingot skin of relatively clean metal low in carbon and other solutes. Such ingots are best suited for the manufacture of steel sheets.

Capped Steel

Production of capped steel is a variation of rimmed steel production practice. The rimming action is allowed to begin normally, but is then terminated after about a minute, by sealing the mold with a cast-iron cap. In steels with carbon content greater than 0.15% the capped ingot practice is usually applied to the production of sheet, strip, wire and bars.

Semi-Killed Steel

Semi-killed steel is deoxidized less than killed steel. This results in the presence of just enough oxygen in the molten steel to react with carbon and form sufficient carbon monoxide to counterbalance the solidification shrinkage. This type of steel generally has a carbon content within the range of 0.15–0.3% and finds wide application in structural shapes.

Killed Steel

Killed steel is deoxidized to such an extent that there is no gas evolution during solidification. Aluminum together with ferro-alloys of manganese and silicon are used for deoxidation. In some applications, calcium silicide or other special strong deoxidizers are also used. In order to minimize piping, almost all killed steels are cast in hot-topped big-end up molds.

Killed steels are generally used when a homogeneous structure is required in the finished product. They include alloy steels, forging steels and steels for carburizing. Steel requiring deep drawing and certain extra-deep-drawing steels, with low-carbon up to a maximum of 0.12% is of this grade. These steels usually have a substantial amount of aluminum added for killing; in the ladle, in the mold, or both.

Deoxidation of steel by aluminum suppresses the formation of carbon monoxide during solidification, which controls formation of blowholes. There are many steel processing operations where aluminum killing of steel is undesirable. Steels that are required for large castings are not killed with aluminum, because the suppression of carbon monoxide gas causes piping defects, and also because inclusion of alumina causes defective castings.

It has been recognized from the early days of continuous-casting operations that casting difficulties and poor surface conditions are often experienced with aluminum-killed steels. Hence, other forms of deoxidation are preferred over aluminum killed steel. The alternative options may include silico-mana-ganese deoxidation and/or vacuum carbon deoxidation.

DEOXIDATION EQUILIBRIA

In the previous paragraphs of this chapter we discussed the importance, type and limitations of deoxidation. Let us now discuss the physics of the process.

Deoxidation reactions can be described using the deoxidation equilib-rium constant. A wide spectrum of deoxidation equilibria pertaining to the most common deoxidants for steel are summarized in Table 1-5-2 below. The table is a log-log plot of the concentration of oxygen in solution in liquid steel against that of the added elements.

In all cases discussed above, the oxygen and the alloying element in solu-tion are in equilibrium with the appropriate gas, liquid or solid oxide phases at 1,600°C, e.g., 1 atm CO, pure B_2O_3, pure Al_2O_3 etc.

Deoxidation reactions can be described using the deoxidation equilibrium constant. The reaction when the alloying element (M) is added to the steel can be represented by Equation (17) below:

$$M_xO_y = xM + Yo \tag{17}$$

TABLE 1-5-2 Solubility of the Products of Deoxidation in Liquid Iron

Equilibrium Constant K*	Composition Range	K at 1600°C	log K
$[a_{Al}]^2[a_O]^4$	<1 ppm Al	1.1×10^{-15}	$-71600/T + 23.28$
$[a_{Al}]^2[a_O]^3$	<1 ppm Al	4.3×10^{-14}	$-62780/T + 20.17$
$[a_B]^2[a_O]^3$		1.3×10^{-8}	
$[a_C] [a_O]^3$	>0.02% C	2.0×10^{-3}	$-1168/T - 2.07$
$[a_{Cr}]^2[a_O]^3$	>3% Cr	1.1×10^{-4}	$-40740/T + 17.78$
$[a_{Mn}] [a_O]$	>1% Mn	5.1×10^{-2}	$-14450/T + 6.43$
$[a_{Si}] [a_O]^2$	>20 ppm Si	2.2×10^{-5}	$-30410/T + 11.59$
$[a_{Ti}] [a_O]^2$	<0.3% Ti	2.8×10^{-6}	
$[a_{Ti}] [a_O]$	>5% Ti	1.9×10^{-3}	
$[a_V]^2[a_O]^4$	<0.10V	8.9×10^{-8}	$-48060/T + 18.61$
$[a_V]^2[a_O]^3$	>0.3% V	2.9×10^{-6}	$-43200/T + 17.52$

Activities are chosen such that $a_{Mn} \equiv$ %Mn and $a_O \equiv$ %O when %M→O
Square brackets [ ] denote component present in molten steel
Temperature (T) is on the Kelvin scale.

The deoxidation constant, assuming that pure M_xO_y forms, is given as Equation (18) below. The term pure M_xO_y describes the unit activity for M_xO_y.

$$K = (h_M)^x(h_O)^y \qquad (18)$$

Where h_M and h_O are the Henrian activities. These are defined as the activity of component that is equal to its weight percent at infinite dilution in iron, as given by Equation (19) below:

$$H_i = f_i(\text{wt.\% } i) \qquad (19)$$

The activity coefficient f_i can be corrected for alloying elements by use of the interaction parameter e^j_i, which is defined by following equation:

$$(d \log f_i /d \log \text{wt}\% j) = e^j_i \qquad (20)$$

Table 1-5-3 shows collected data for carbon steel and stainless steel, showing the coefficients of interaction for the common elements of carbon and stainless steels at 1,600°C (2,912°F).

The activity coefficient for most low alloy steels encountered in ladle metallurgy is taken as unity, and Equation (18) reduces to the following:

$$K_M = (\%M)^x(\%O)^y \qquad (21)$$

To illustrate how to use these constants, let's consider a steel that contains silicon [0.1%] at 1,600°C (2,912°F) that is in equilibrium with SiO_2. In this case the value of K_{Si} is given by following equation:

$$K_{Si} = (\%Si)(\%O)^2 \qquad (22)$$

$K_{Si} = 2.2 \times 10^{-5}$
    Therefore:

$$(\%O)^2 = 2.2 \times 10^{-4}$$
$$(\%O) \approx 0.015 \text{ or } 150\,\text{ppm.}$$

It is important to note that these calculations are for the soluble oxygen content; the total oxygen content, including both the soluble oxygen and the oxygen associated with inclusions could be much higher.

For single element deoxidation, the solubility of oxygen in liquid iron at 1,600°C (2,912°F) is given as a function of the concentration of the alloying element. In each case, the melt is in equilibrium with the respective pure oxide, e.g., SiO_2, Al_2O_3. It can be clearly seen that aluminum is the strongest of the common deoxidizers followed by titanium. Rare earths are equally strong deoxidizers as aluminum, but their use is limited to control radiation.

TABLE 1-5-3 The Coefficients of Interaction for the Common Elements of Carbon and Stainless Steels at 1600°C (2912°F)

Metal		Al	C	Mn	P	S	Si	Ti	H	N	O	Cr	Ni
Carbon steel	$\%_i$		0.05	0.45	0.02	0.01	0.3	0.05					
	f_i	1.05	1.06	1.0	1.1	1.0	1.15	0.93	1.0	0.97	0.85		
	a_i		0.053	0.45	0.022	0.01	0.345	0.046					
Stainless steel	$\%_i$		0.05	0.45	0.02	0.01	0.3	0.05				18	8
	f_i	3.6	0.49	1.0	0.32	0.66	1.24	9.4	0.93	0.17	0.21	0.97	1.0
	a_i		0.025	0.45	0.006	0.007	0.372	0.47				17.5	8.0

THE IRON-IRON CARBIDE PHASE DIAGRAM

Iron is an allotropic metal – it can exist in more than one type of lattice structure depending on the temperature. The temperature at which the structural changes occur depends on the alloying elements which are present in iron, especially the carbon content.

The iron-iron carbide phase diagram that was discussed in the previous chapter is a good basis for understanding the effect of temperature and components on the properties of steel. In doing so, the portion of the diagram between pure iron and interstitial Fe_3C (iron carbide) is of significance.

Classification of Steels

Steel is classified by a variety of different systems, depending on the following parameters:

- The composition, such as carbon, low-alloy or stainless steel.
- The manufacturing methods, such as open hearth, basic oxygen process, or electric furnace methods.
- The finishing method, such as hot or cold rolling.
- The product form, such as bar plate, sheet, strip, tubing or structural shape.
- The deoxidation practice, such as killed, semi-killed, capped or rimmed steel.
- The microstructure, such as ferritic, pearlitic and martensitic.
- The required strength level, as specified by various industry standards such as API, ASTM/ASME etc.
- The heat treatment, such as annealing, quenching and tempering, and thermomechanical processing etc.
- Quality descriptors, such as forging quality and commercial quality.

AISI-SAE Designation	Type of Steel with Typical Grades	Nominal Alloy Content (%)		
Carbon Steels				
10xx	Plain carbon steel: 1005, 1010, 1016, 1030 etc.	Manganese up to 1% max		
11xx	Resulfurized: 1110, 1117, 1137 etc.			
12xx	Resulfurized and Rephosphorized: 1211, 1212, 1213, 1215, 12L14 (this grade include up to 0.35% lead)			
15xx	Plain carbon steel: 1513, 1522, 1526, 1548,1561 and 1566 etc.	Manganese range from 1% to 1.65%		
Manganese Steels				
13xx		Manganese 1.75%		
Nickel Steels				
23xx		Nickel 3.5		
		Nickel 5		
Nickel Chromium Steels				
31xx		Nickel 1.25, Cr 0.65and 0.80		
32xx		Nickel 1.75, Cr 1.07		
33xx		Nickel 3.50, Cr 1.5 and 1.57		
34xx		Nickel 3.00, Cr 0.77		
Molybdenum Steels				
40xx	4023, 4024, 4027, 4028 etc.	Mo 0.20 and 0.25		
44xx		Mo 0.40 and 0.52		
Chromium Molybdenum Steels				
41xx			Cr	Mo
	4118		0.50	0.12
	4130, 4137, 4140,		0.80	0.20
	4142, 4145, 4147, 4150		0.95	0.25
	4161		0.70-0.90	0.30
Nickel Chromium Molybdenum Steels				
43xx	4320,	Ni	Cr	Mo
		1.65 to 2.00	0.40 to 0.60	0.20 to 0.30
	4340,	1.65 to 2.00	0.70 to 0.90	0.20 to 0.30
47xx	4720	0.90 to 1.20	0.35 to 0.55	0.15 to 0.25
81xx		0.30	0.40	0.12
86xx		0.55	0.50	0.20
87xx		0.55	0.50	0.25
88xx		0.55	0.50	0.35
93xx		3.25	1.20	0.12
94xx		0.45	0.40	0.12
97xx		0.55	0.20	0.20
98xx		1.00	0.80	0.25
Nickel Molybdenum Steels				
46xx		Ni		Mo
		0.85		0.20
		1.82		0.25
48xx		3.50		0.25
Chromium Steels				
50xx		Cr 0.27, 0.40, 0.50, 0.65		
51xx		Cr 0.80, 0.87, 0.92, 1.00, 1.05		
52xx		Cr 0.50, Carbon 1.00		

FIGURE 1-6-1   Classification of Steel.

	Chromium Vanadium Steels		
61xx		Cr	V
		0.60,0.80,0.95	0.10, 0.15
	Chromium Tungsten Steels		
72xx		Cr 0.75; W 1.75	
	Silicon Manganese Steels		
92xx		Si: 1.40, 2.00; Mn 0.65, 0.82, 0.85; Cr 0.65	
	High-Strength Low-Alloy (HSLA) Steels		
9xx		Various SAE grades	
xxBxx		B indicates added Boron	
xxLxx		L indicates Lead addition	
	XX indicates steel designation for carbon steel and low alloys.		
	Stainless Steels		
AISI designation	SAE designation		
2xx	302xx	Chromium Manganese and	

FIGURE 1-6-1   (Continued)

CARBON STEELS

The American Iron and Steel Institute (AISI) defines carbon steel as the steel that has no minimum specified content or requirements for chromium, cobalt, niobium, molybdenum, nickel, titanium, tungsten, vanadium or zirconium, or specified requirement for any other element to be added specifically to obtain a desired alloying effect; in such steel the specified copper should be between 0.40 and 0.60%, the maximum limit for manganese is 1.65%, and that for silicon is 0.60%.

Carbon steel is classified according to deoxidation practices, as rimmed steel, capped steel, killed steel or semi-killed steel. We have discussed these types of steel in Chapter 5.

Steels are also classified by their carbon content. Carbon steels containing up to 2% total alloying elements are further divided into low-carbon steels, medium-carbon steels, high-carbon steels, and ultrahigh-carbon steels; each of these classification is discussed below.

Low-Carbon

Low-carbon steels contain up to 0.30% carbon. The majority of this class of steel consists of flat-rolled products like sheet or strip, and usually they are in the cold-rolled and annealed condition. These steels have high-formability, as they contain very low carbon – usually less than 0.10% C, with up to 0.4% Mn.

For rolled steel structural plates and sections, the carbon content is often increased to approximately 0.30%, and the manganese content to 1.5%. These materials are useful for stampings, forgings, seamless tubes, and as boilerplates.

Medium-Carbon

Medium-carbon steels are similar to low-carbon steels, except that they have a carbon content between 0.30 and 0.60%, and manganese between 0.60 and 1.65%. Increasing the carbon content to approximately 0.5% with an accompanying increase in manganese allows medium carbon steels to be used in the quenched and tempered condition. These steels are mainly used for making shafts, axles, gears, crankshafts, couplings and forgings. Steels with carbon levels ranging from 0.40 to 0.60% are used for rails, railway wheels and rail axles.

High-Carbon

High-carbon steels contain carbon at levels of 0.60 to 1.00%; the manganese content ranges from 0.30 to 0.90%. High-carbon steels are used for some hand tools, spring materials, high-strength wires etc.

Ultrahigh-Carbon

Ultrahigh-carbon steels are often experimental alloys containing 1.25 to 2.0% carbon. They are often thermomechanically processed to produce consistent and ultra-fine microstructures; that is equi-axial grains of spherical, discontinuous pro-eutectoid carbide particles.

HIGH-STRENGTH LOW-ALLOY STEELS (HSLA)

High-strength low-alloy (HSLA) steels are micro-alloyed steels; they are designed to provide better mechanical properties and may also have greater resistance to atmospheric corrosion than conventional carbon steels.

HSLA steels have a low carbon content (0.05–0.25%), in order to produce adequate formability and weldability, and they have manganese contents up to 2.0%. Small quantities of chromium, nickel, molybdenum, copper, nitrogen, vanadium, niobium, titanium and zirconium are added in various combinations to impart specific properties.

Classification of HSLA

Weathering steels are designed to exhibit superior atmospheric corrosion resistance.

Control-rolled steels are hot rolled in accordance with a predetermined rolling schedule. These steels are designed to develop a highly deformed austenite structure that on cooling transforms to a very fine equi-axial ferrite structure.

Pearlite-reduced steels are strengthened by very fine-grain ferrite and precipitation hardening, but have a low carbon content and therefore little or no pearlite in the microstructure.

Microalloyed steels have elements such as niobium, vanadium, and titanium added at very low levels, for refinement of grain size or precipitation hardening. Often the combined total of these alloys is limited, as well as their individual levels being very small.

Acicular ferrite steel has very low carbon content and develops sufficient hardenability on cooling to transform to a very fine, high-strength, acicular, ferrite structure rather than the usual polygonal ferrite structure.

HSLA steels may also contain small amounts of calcium, rare earth elements, or zirconium for sulfide inclusion shape control.

LOW-ALLOY STEELS

Low-alloy steels constitute a category of ferrous materials that exhibit mechanical properties superior to plain carbon steels, due to the addition of alloying elements such as nickel, chromium, and molybdenum. Total alloy content can range from 2.07% up to levels just below that of stainless steels, which contain a minimum of 10% chromium.

For many low-alloy steels, the primary function of the alloying elements is to increase hardenability in order to optimize mechanical properties and toughness after heat treatment. In some cases, however, alloy additions are to reduce environmental degradation under certain specified service conditions. As with steels in general, low-alloy steels can be classified according to their **chemical composition,** such as nickel steels, nickel-chromium steels, molybdenum steels, and chromium-molybdenum steels.

They are also classified on the basis of their **heat treatment,** such as quenched and tempered, normalized and tempered, or annealed.

Because of the wide variety of possible chemical compositions, and the fact that some steels are used in more than one heat-treated condition, some overlap exists among the alloy steel classifications. Four major groups of alloy steels are identified and discussed here.

Low-Carbon Quenched and Tempered Steels

These steels combine high yield strength, ranging from 50 to 150 ksi (350 to 1,035 MPa) with good notch toughness, ductility, corrosion resistance, and weldability. There are different combinations of these steels with varying characteristics based on their intended applications.

Medium-Carbon Ultrahigh-Strength Steels

These are structural steels with yield strengths that can exceed 200 ksi (1,380 MPa). Many of these steels are covered by SAE/AISI designations, or are proprietary compositions. Product forms include billet, bar, rod, forgings, sheet, tubing, and welding wire.

Bearing Steels

Bearing steels, as the name suggests, are used for ball and roller bearing applications. They contain low carbon, of about 0.10 to 0.20%, and these steels are casehardened. Many of these steels are covered by SAE/AISI designations.

Chromium-Molybdenum Heat-Resistant Steels

Chromium-molybdenum heat-resistant steels contain 0.5 to 9% chromium and 0.5 to 1.0% molybdenum. The carbon content is usually below 0.2%. The chromium provides improved oxidation and corrosion resistance, and the molybdenum increases strength at elevated temperatures. They are generally supplied in the normalized and tempered, quenched and tempered or annealed condition. Chromium-molybdenum steels are widely used in the oil and gas industries, and in fossil fuel and nuclear power plants.

AISI SERIES

As stated earlier, steels may be classified by method of manufacture, use, or by chemical composition. Manufacturing method categories include Bessemer steel, open-hearth steel, electric-furnace steel, crucible steel, etc. Some steel specifications specify the particular method required to comply with a standard. Classifications based on use include machine steel, spring steel, boiler steel, structural steel, or tool steel.

Classification by chemical composition is the most common method. Steel specification systems have been generated by AISI and SAE in America that use an elaborate numbering system to indicate the approximate alloy content.

Sometimes letter prefixes are included to designate the steel-making process (B=acid Bessemer, C=basic open-hearth, E=basic electric-furnace). The last two numbers indicate the carbon content, for example AISI 1020 steel contains 0.20% carbon. The following is a list of some AISI steels and their general descriptions. For more detailed information about any of these steels, refer to AISI publications.

Some Examples of AISI Classifications

- 10xx Basic open-hearth and acid Bessemer carbon steels
- 11xx Basic open-hearth and acid Bessemer carbon steels, high sulfur, low phosphorus
- 12xx Basic open-hearth carbon steels, high sulfur, high phosphorus
- 13xx Manganese 1.75%
- 40xx Molybdenum 0.20% or 0.25%
- 41xx Chromium 0.50%, 0.80%, or 0.95%, molybdenum 0.12%, 0.20%, or 0.30%
- 43xx Nickel 1.83%, chromium 0.50%, or 0.80%, molybdenum 0.25%
- 46xx Nickel 0.85% or 1.83%, molybdenum 0.20% or 0.25%
- 47xx Nickel 1.05%, chromium 0.45%, molybdenum 0.20% or 0.35%
- 48xx Nickel 3.5%, molybdenum 0.25%
- 51xx Chromium 0.80%, 0.88%, 0.93%, 0.95%, or 1.00%
- 5xxxx Carbon 1.4%, chromium 1.03% or 1.45%
- 61xx Chromium 0.60% or 0.95%, vanadium 0.13% or 0.15%
- 86xx Nickel 0.55%, chromium 0.50%, molybdenum 0.20%
- 87xx Nickel 0.55%, chromium 0.50%, molybdenum 0.25%
- 88xx Nickel 0.55%, chromium 0.50%, molybdenum 0.35%
- 92xx Silicon 2.00%

Cast Iron

Chapter Outline

The term 'cast iron' covers a family of ferrous alloys. It is an iron alloy that contains more than 2% carbon, and silicon in the range 1 to 3%. The extent of alloying and associated control of the heat and cooling process can give a very wide variety of properties to different grades of cast iron. As we saw in the iron carbon phase diagram in chapter 4, Figure 1-4-4, the correct position of cast iron is indicated to the right of steel at 2.1% carbon.

TYPES OF CAST IRON

Cast iron is used in the oil and gas industry for making water conduit pipes, packer parts, boilers, valve bodies and valve parts. Cast iron is the form of material that is favored by foundry-men over cast steel for the following reasons:

1. Ease of production by Cupola furnace.
2. Lower melting temperature than steel.
3. Excellent fluidity of molten cast iron.

The construction of a cupola is somewhat like a blast furnace; however, apart from its very different use, it is much smaller and less costly to operate. The furnace is charged with coke, pig iron scrap metal, and limestone as a flux and slag former. Other elements are introduced into the melt through the addition of ferroalloys. Cast irons have low ductility, hence they cannot be rolled, drawn, or worked on at room temperature. Most cast irons have lower strength than most steels. Cast irons are generally classified according to their microstructure, which depends on the carbon content, the alloy and impurity content, the cooling rate during and after freezing, and heat treatment after casting. The condition and physical form of the carbon controls the properties of cast iron.

The composition of cast iron varies significantly, depending upon the grade of pig iron, scrap used but in general it contains carbon in the range of about 2–4% by weight. The other alloying elements are manganese (Mn), phosphorus (P), sulfur (S) and high silicon concentrations, together with a much greater concentration of impurities than steels.

The mode and concentration of carbon in the cast iron is controlled to produce White Cast Iron, Gray Cast Iron, or Malleable Cast Iron. These forms of cast iron differ significantly from each other in their mechanical properties and weldability.

Both carbon and silicon levels influence the nature of iron castings. A factor called the Carbon Equivalent (CE) approximates their impact on solidification. The impact of two elements can be determined by following relationship:

$$CE = C + (Si/3)$$

For a more accurate assessment, the effect of phosphorus is also considered, and the formula is suitably modified:

$$CE = C + (Si + P/3)$$

The comparison with iron and carbon (Fe-C) eutectic system (i.e. 4.3% C) and a knowledge of the CE of the casting allows us to predict whether the resulting cast iron will behave as a hypoeutectic or hypereutectic on solidification. If the CE value of the given casting is closer to the eutectic value of the casting, the liquid state will persist to a relatively low temperature and solidification would take place over a small temperature range. This is important, because it promotes uniform properties in the casting.

In hypereutectic irons, where the CE is greater than the eutectic value of 4.3%, the tendency to form kish-graphite increases. The kish-graphite is the proeutectic graphite that freely floats within the molten iron, and precipitates on solidification under normal cooling conditions. In such iron, lower CE values increase the chances of formation of white or mottled cast iron on solidification.

The CE of a cast iron also helps to distinguish the gray irons, which cool into a microstructure containing graphite, from the white irons where the carbon is present mainly as cementite.

A high cooling rate and a low CE favors the formation of white cast iron, whereas a low cooling rate or a high CE promotes gray cast iron.

During solidification, the major proportion of the carbon precipitates in the form of graphite or cementite. When solidification is just complete, the precipitated phase is embedded in a matrix of austenite that has an equilibrium carbon concentration of about 2% by weight. On further cooling, the carbon concentration of the austenite decreases as more cementite or graphite precipitates from the solid solution. For conventional cast irons, the austenite then decomposes into pearlite at the eutectoid temperature. However, in gray cast irons, if the cooling rate through the eutectoid temperature is sufficiently slow, then a completely ferritic matrix is obtained, with the excess carbon being deposited on the already existing graphite.

WHITE CAST IRON

White cast iron is formed when the carbon in solution is not able to form graphite on solidification. White cast irons are hard and brittle; they cannot easily be machined. They are unique in that they are the only member of the cast iron family in which carbon is present only as a carbide. White cast iron has a light appearance due to the absence of graphite. It has a high compressive strength and retains good hardness and strength at higher temperatures. The presence of different carbides, depending on the alloy content, makes white cast irons extremely hard and abrasion resistant but very brittle. An improved form of white cast iron is chilled cast iron, also discussed in this chapter.

The microstructure of white cast iron contains cementite (white) and pearlite. It contains interdendritic cementite (white) which sometimes has a Widmanstiitten ('spiky') appearance. Austenite forms as the proeutectic constituent before the eutectic reaction (liquid transforms to austenite and cementite), and later transforms to pearlite and cementite upon cooling below the eutectoid temperature, about 723°C (1,333°F). White cast iron is named after its white surface when fractured, which is caused by its carbide impurities, allowing cracks to pass straight through.

MALLEABLE CAST IRON

Also referred to as malleable iron, the carbon in this class of cast iron is present as irregularly shaped graphite nodules instead of flakes. It is essentially a white cast iron that has been heat treated to convert the iron carbide into the irregularly shaped graphite nodules. This form of graphite is called temper carbon because it is formed in the solid state during the heat treatment.

Although the cementite found in cast iron is considered to be metastable, heat treatment can cause it to transform into iron and carbon. Malleabilization is performed to convert all of the combined carbon in white iron into irregular nodules of graphite and ferrite. Two annealing stages are performed at about 899°C (1,650°F) to 927°C (1,700°F) for up to 72 hours. A tough ferritic matrix surrounds the resultant temper carbon so that malleable cast iron has higher strength and ductility than gray cast iron.

Malleable iron is classified into three grades:

1. Ferritic malleable
2. Pearlite malleable
3. Martensitic malleable

We shall now briefly discuss each of these types.

FERRITIC MALLEABLE IRON

This category of malleable iron has two types; white heart iron and black heart iron.

White Heart Cast Iron

White heart cast iron is essentially cast as white cast iron and made malleable by annealing in an oxidizing medium. During the heat treatment substantial decarburization occurs, and the remaining carbon is precipitated in the form of graphite nodules. This type of iron normally contains about 0.6 to 1.3% silicon, which is just enough to promote cementite decomposition during the heat treatment, but not enough to produce graphite flakes during casting. White heart malleable iron is made by heating to a temperature of 900°C (1,650°F) in an oxidizing atmosphere to remove carbon from the surface of white iron castings.

Black Heart Cast Iron

Black heart cast iron has a matrix of ferrite with interspersed nodules of temper carbon. In the United States the term 'Cupola malleable iron' is used for black heart malleable iron, as it is produced from a Cupola furnace. The strength and ductility of this grade is very low, hence it is not favored as structural material.

PEARLITE MALLEABLE CAST IRON

Pearlite malleable cast iron has combined carbon in the matrix; this results in higher strength and hardness than ferritic malleable iron.

The production process involves controlling the carbon content and heat treatment cycle. Some carbon can be retained as finely distributed iron carbide. This combined carbon increases the strength and hardness of the cast iron.

MARTENSITIC MALLEABLE IRON

Martensitic malleable iron is initially produced as pearlitic iron. It is subsequently quenched and tempered to make martensitic malleable iron.

GRAY CAST IRON

Gray cast iron is a broad term used for a number of cast irons whose microstructure is characterized by the presence of flake graphite in the ferrous matrix. Such castings often contain 2.5 to 4% carbon, 1 to 3% silicon and some manganese, ranging from 0.1 to 1.2%.

This is one of the most widely used alloys of iron. The strength of gray cast iron depends on the matrix in which graphite (free carbon) is embedded. The matrix can range from ferrite to pearlite and various combinations of the two phases. Large graphite flakes reduce strength and ductility, so inoculants are used to promote fine flakes.

Gray cast iron is named after its gray fractured surface, which occurs because the graphitic flakes deflect a passing crack and initiate countless new cracks as the material breaks. White cast irons are hard and brittle. Gray cast irons are softer, with a microstructure of graphite in transformed-austenite and cementite matrix. The graphite flakes, which are rosettes in three dimensions, have a low density and hence compensate for the freezing contraction, thus giving good castings that are free from porosity.

The flakes of graphite have good damping characteristics and good machinability, because the graphite acts as a chip-breaker and lubricates the cutting tools. In applications involving wear, the graphite is beneficial because it helps retain lubricants. However, the flakes of graphite are also stress concentrators, leading to poor toughness. The recommended applied tensile stress is therefore only a quarter of its actual ultimate tensile strength.

The presence of sulfur in cast irons is known to favor the formation of graphite flakes. The graphite can be induced to precipitate in a spheroidal shape by removing the sulfur from the melt using a small quantity of calcium carbide. This is followed by addition of minute amounts of magnesium or cerium, which poisons the preferred growth directions and leads to isotropic growth, resulting in spheroids of graphite. The calcium treatment is necessary before the addition of magnesium, since the latter also has an affinity for both sulfur and oxygen, whereas its spheroidizing ability depends on its presence in solution in the liquid iron. The magnesium is frequently added as an alloy with iron and silicon (Fe-Si-Mg), rather than as pure magnesium. However, magnesium tends to encourage the precipitation of cementite, so silicon in the form of ferro-silicon is also added to ensure the precipitation of carbon as graphite. The ferro-silicon is the *inoculant* in the system.

Gray cast iron is by far the oldest and most common form of cast iron. As a result, it is assumed by many to be the only form of cast iron, and the terms 'cast iron' and 'gray iron' are used interchangeably. Unfortunately the only commonly known property of gray iron – brittleness – is also assigned to 'cast iron' and hence to all cast irons. Gray iron is so named because its fracture has a gray appearance. It contains carbon in the form of flake graphite in a matrix which consists of ferrite, pearlite or a mixture of the two.

The fluidity of liquid gray iron, and its expansion during solidification due to the formation of graphite, has made this metal ideal for the economical production of shrinkage-free, intricate castings such as engine blocks.

The flake-like shape of graphite in gray iron exerts a dominant influence on its mechanical properties. The graphite flakes act as stress raisers which may prematurely cause localized plastic flow at low stresses, and initiate fracture in the matrix at higher stresses. As a result, gray iron exhibits no elastic behavior and fails in tension without significant plastic deformation, but has excellent damping characteristics. The presence of graphite flakes also gives gray iron excellent machinability and self-lubricating properties.

ASTM Specification A-48 lists several classes of gray cast irons, based on tensile strength. A range from 20 to 60 ksi tensile strength is identified. Cast irons above 40 ksi tensile strength are considered high-strength irons. Compressive strength is much more important than tensile strength for many applications, and in many cases gray cast iron performs better than steel in compression loading applications.

Castability of Gray Cast Iron

Two important parameters for a successful casting are the fluidity of molten metal and the cooling rate. These are also affected by thickness and variations in thickness of the section of casting. The term 'section sensitivity', when used to define the castability, is an attempt to correlate properties in the critical sections with composition and cooling rate.

The fluidity of the molten iron is affected by the mould conditions, pouring rate, and amount of super heat above the freezing (liquidus) temperature. It may be noted that as the total carbon (TC) decreases, the liquidus temperature increases and the fluidity at given pouring temperature also decreases.

FIGURE 1-7-1 Microstructure of gray cast iron, Fe, 3.2%C and Si 2.5% by wt, containing graphite flakes in a matrix, which is pearlitic. The speckled white regions represent a phosphide eutectic. *Picture courtesy of Bombay Malleable, Andheri-Kurla Road, Bombay (1983).*

CHILLED CAST IRON

When a localized area of a gray cast iron is cooled very rapidly from the melt, cast iron is formed. This type of white cast iron is called chilled cast iron. A chilled iron casting can be produced by adjusting the carbon composition of the white cast iron, so that the normal cooling rate at the surface is just fast enough to produce white cast iron while the slower cooling rate below the surface will produce gray iron. The depth of chill decreases and the hardness of the chilled zone increases with increasing carbon content.

Chromium in small amounts is often used to control chill depth, because of the formation of chromium carbides. Chromium is used at 1 to 4% in chilled iron to increase hardness and improve abrasion resistance. It also stabilizes carbide and suppresses the formation of graphite in heavy sections. When added in amounts of 12 to 35%, chromium imparts resistance to corrosion and oxidation at elevated temperatures.

NODULAR (SPHEROIDAL GRAPHITE) CAST IRON

Nodular iron is also called ductile iron. The graphite is present as tiny balls or spheroids. Since the spheroids interrupt the matrix much less than graphite flakes, nodular cast iron has higher strength and toughness than gray cast iron. The formation of nodules or spheroids occurs when eutectic graphite separates from the molten iron during solidification. The separation of graphite in nodular form is similar to the separation of graphite in gray cast iron, except that the additives facilitate the graphite to take this nodular shape.

Spheroidal graphite (SG) cast iron has excellent toughness; it has higher elongation and is used widely, for example in crankshafts. Unlike malleable iron, nodular iron is produced directly from the melt and does not require heat treatment. Magnesium or cerium is added to the ladle just before casting. The matrix can be either ferrite, pearlite or austenite. The quality of SG iron is excellent and X-ray quality castings are regularly produced.

The latest breakthrough in cast irons is where the matrix of SG cast iron is not pearlite, but bainite. The chemical composition of the SG cast iron is similar to that of the gray cast iron, but with an additional 0.05 wt% of magnesium. This results in a major improvement in toughness and strength. The bainite is obtained by isothermal transformation of the austenite at temperatures below that at which pearlite forms. The process of graphitization is discussed in further detail in the chapter on heat treatment.

Castability, Solidification and Shrinkage

The material as a liquid has high fluidity and excellent castability, but it also has very high surface tension, which means that the sand and molding equipment used must be rigid, high density and be capable of good heat transfer.

The formation of graphite during solidification causes an increase in volume; this is capable of countering the loss in volume due to the liquid-to-solid phase change in metallic constituents. SG iron casings typically require very few risers to work as reservoirs of the molten metal to fill the mold. Because of this property, the ratio of weight of the usable casting to the weight of the metal poured, called 'mold yield' is much higher than for either steel or malleable iron castings. Yet this advantage is not as high as with gray cast iron.

Casting design engineers must compensate for the shrinkage of cast iron. Shrinkages occur during both solidification and subsequent cooling to room temperature, and it is compensated for by designing a pattern that is larger in size than the final produced casting. The shrinkage allowance for ductile and SG iron is 0 to 0.75%, compared to 1% for gray iron, and 2% for white cast iron, and carbon steel castings. The shrinkage allowance for alloy steel castings is 2.5%.

ALLOY CAST IRONS

Alloy cast irons contain specially added elements to modify the physical or mechanical properties. Chromium forms complex iron-chromium carbides that increase strength, hardness, depth of chill, and resistance to wear and heat. Copper breaks up massive cementite and strengthens the matrix. Molybdenum improves mechanical properties such as fatigue strength, tensile strength, heat resistance, and hardness. Nickel is used to control the microstructure by favoring the formation of pearlite.

Stainless Steels

Stainless steels are iron base alloys that contain a minimum of approximately 11% chromium (Cr). This is important for creating a passivating layer of chromium-rich oxide to prevent rusting on the surface. Several stainless steel grades are produced to address specific environmental demands that they are expected to withstand. For this purpose other elements are also added to the steel, including nickel, copper, titanium, aluminum, silicon, molybdenum, niobium, nitrogen sulfur, and selenium. The corrosion resistance and corrosion properties of stainless steel are briefly discussed in this chapter. Readers wanting more in-depth knowledge about these topics are guided to the book 'Corrosion and Corrosion Prevention' by the same author.

STAINLESS STEEL PRODUCTION

Stainless steel production involves a series of processes. First, the steel is melted in an electric arc furnace (EAF), and alloying elements like chromium,

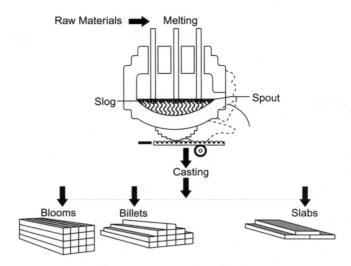

FIGURE 1-8-1   Schematic, stainless steel production.

silicon, nickel, etc. are added and melted in intense heat. Usually this process continues for about 8 to 12 hours. This is followed by casting the molten metal into one of several shapes. These shapes can include blooms, rectangular shapes, billets, round or square shapes of 1.5 inches, or slabs, rods, and tubes.

Forming

The semi-finished steel goes through forming operations, beginning with hot rolling, in which the steel is heated and passed through a series of rolls where the blooms and billets are formed into bars and wires:

- Bars are available in all grades and come in rounds, squares, octagons, or hexagons of 0.25 inches (6 millimeters) in size
- Wire is usually available up to 0.5 inches (13 millimeters) in diameter
- Slabs are formed into plates, strips, and sheets
- Plates are defined as rectangular shapes of more than 0.1875 inches (5 millimeters) thick and over 10 inches (250 millimeters) wide
- Strips are defined as rectangular shapes of less than 0.185 inches (5 millimeters) thick and less than 24 inches (610 millimeters) wide
- Sheets are defined as rectangular shapes of less than 0.1875 inches (5 millimeters) thick and more than 24 inches (610 millimeters) wide.

Heat Treatment

Once the final shape is formed, most types of stainless steel must go through an annealing process. Annealing is a heat treatment in which the steel is heated

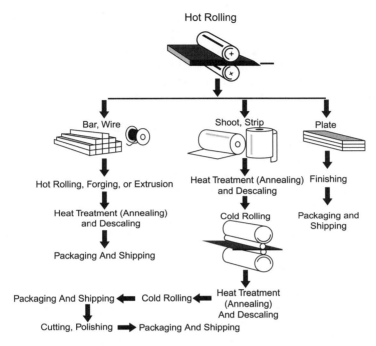

FIGURE 1-8-2   Schematic, Stainless steel finishing process.

and cooled under controlled conditions to relieve internal stresses and soften the metal. Some steels are heat treated for higher strength. However, such a heat treatment, known as age hardening, requires careful control, as even small changes from the recommended temperature, time, or cooling rate can seriously affect the end properties. Lower aging temperatures produce high strength with low fracture toughness, while higher-temperature aging produces a lower strength, tougher material.

Though the rate of heating to reach the aging temperature of 900 to 1 000°F (482 to 537°C) does not affect the properties, the control on cooling rate is very important as it does affect the properties. A post-aging quenching treatment is often carried out to increase the toughness of the steel. One such process involves water quenching, in which the material is quenched in an ice water bath at about 1.6°C (35°F) for two hours.

Different type of steels such as austenitic, ferritic, or martensitic receive different heat treatment. Austenitic steels are heated to above 1 037°C (1 900°F) for a time depending on the thickness. Thick sections are quenched in water and thinner sections are either cooled by air cooling or air blasting. Rate of cooling is of importance, if cooled too slowly carbide precipitation can occur.

The carbide precipitation can be eliminated by thermal stabilization. In this method, the steel is held at 815 to 870°C (1 500 to 1 600°F) for several hours.

Descaling. Heat treatment causes a scale to build up on the surface of the steel. The scale is removed by several methods.

Pickling is done to remove scale, the material is immersed in a bath of nitric-hydrofluoric acid and the acid leaches out the scale from the surface.

Electrocleaning is also carried out to descale the steel. The process involves application of an electric current to the surface using a cathode and phosphoric acid.

The annealing and descaling are carried out at different stages depending on the type of steel being made. For example, bar and wire go through further forming steps (more hot rolling, forging, or extruding) after the initial hot rolling before being annealed and descaled. Sheet and strip, on the other hand, go through an initial annealing and descaling step immediately after hot rolling. After cold rolling (passing through rolls at a relatively low temperature), which produces a further reduction in thickness, sheet and strip are annealed and descaled again. A final cold rolling step prepares the steel for final processing.

Cutting Stainless Steel

A cutting operation is usually necessary to obtain the desired blank shape or size. This is done to trim the part to final size. Mechanical cutting is accomplished by a variety of methods, including straight shearing by guillotine knives, circle shearing by circular knives horizontally and vertically positioned, and blanking by metal punches and dies to punch out the shape by shearing. Nibbling is a process used for cutting by blanking out a series of overlapping holes and is ideally suited for irregular shapes, as only some stainless steels can be saw cut by high speed steel blades.

Some stainless steel can also be cut using flame cutting, which involves a flame-fired torch using oxygen and propane in conjunction with iron powder. This method is clean and fast.

Another cutting method that is used is plasma jet cutting. To make a cut this process uses an ionized gas column in conjunction with an electric arc passing through a small orifice, where the force of the gas and high heat generated by the gas plasma melts the metal and makes the cut.

Finishing

Surface finish is an important requirement for stainless steel products, depending on the end application. The surface finish is a very important property to specify.

The main reasons to consider specifying the surface finish could include one or all of the following.

1. The appearance.
2. Process convenience.

3. Corrosion protection.
4. To facilitate lubrication – often rougher surface is specified.
5. Surface condition specific to facilitate further manufacturing steps.

FABRICATION OF STAINLESS STEEL

After the stainless steel in its various forms is packed and shipped to the fabricator or end user, a variety of secondary processes are needed to make it useful for specific services. Further shaping is accomplished using a variety of secondary processing that may include rolling forming, press forming, forging, press drawing, and extrusion, welding cutting, additional heat treating, machining, and cleaning processes.

WELDING AND JOINING

The details of welding stainless steel are discussed in Section 2 of this book, however a brief introduction is included here.

There are a variety of methods for joining stainless steel, welding being the most common. Several variations of fusion and resistance welding processes can be used. In fusion welding, heat is provided by an electric arc struck between an electrode and the metal to be welded. In resistance welding, bonding is the result of heat and pressure. Heat is produced by the resistance to the flow of electric current through the parts to be welded, and pressure is applied by the electrodes. After parts are welded together, they must be cleaned around the joined area.

TYPES OF STAINLESS STEELS

Stainless steels are used for both corrosion and high-temperature resistance applications. A three-number system as listed below is used to identify stainless steels. In the following list the three digit number generally corresponds with the type and group of alloying elements. The first digit is the group identifier and the last two digits identify the specific type of alloy:

- 2xx Chromium-nickel-manganese non-hardenable, austenitic, nonmagnetic
- 3xx Chromium-nickel non-hardenable, austenitic, nonmagnetic
- 4xx Chromium hardenable, martensitic, magnetic
- 4xx Chromium non-hardenable, ferritic, magnetic
- 5xx Chromium low chromium, heat-resisting.

The corrosion resistance in stainless steel is obtained by the presence of a thin, adherent, stable chromium oxide or nickel oxide film, which protects the steel. A minimum of 10% Cr is required to develop this tenaciously adhering, uniformly developed oxide film.

CLASSIFICATION OF STAINLESS STEEL

Stainless steels are classified into five families. Four of these are based on their crystallographic structures, these are martensitic, ferritic, austenitic, and austenitic plus ferritic (called duplex). The fifth class is based on the type of heat treatment used to produce certain properties. These are called precipitation-hardenable alloys, often also called PH steels. Each of these are discussed further in the text.

Martensitic Stainless Steels

These are essentially FeCrC alloys that have boy centered tetragonal (bct) crystalline structure, which is martensite in a hardened condition. These steels are ferromagnetic; they can be hardened by heat treatment. These steels can resist corrosion in mild environments. Addition of a small amount of nickel improves corrosion resistance properties. Typically, these steels contain 11.5% to 18% chromium and up to 1.2% carbon. The chromium to carbon ratio is balanced to maintain martensitic structure. Examples of this group of steel are Types 403, 410, 420, and 501.

Other elements like vanadium, silicon, tungsten and niobium are also added to induce tempering after hardening. The alloying is in very limited scale to maintain the martensitic structure.

Properties of Martensitic Stainless Steel

In annealed condition these steels exhibit good tensile properties. Generally these steels have yield strength of approximately 275 MPa (40 ksi). Steel responds to cold work by moderate increase in hardness. In annealed condition these alloys have good toughness and ductility, which decreases as hardness is increased by heat treatment. Heat treatment hardening and tempering increases yield strength up to 1 900 MPa (275 ksi). Typically the hardness values range from 150 HB in annealed condition to 600 HB in fully hardened condition.

Martensitic stainless steels are magnetic and can be easily cold-worked and machined. In annealed condition they have good toughness, moderate corrosion resistance, and are easily hot-worked. They are most corrosion resistant when properly heat-treated. Because of the high alloy content, these steels undergo sluggish transformations and this results in high hardenability. Maximum hardness can be achieved by air-cooling. Type 416 steels contain a small amount of sulfur to improve machinability, which also reduces corrosion resistance.

Ferritic Stainless Steels

These are straight iron-chromium alloys that contain chromium in the range of 14 to 27%. Examples of this group of steel are Type 405 and Type 430 steel.

This type of steel has a bcc structure. To obtain specific properties, some grades are alloyed with molybdenum, aluminum, silicon, titanium and niobium.

These steels are ferromagnetic, they posses good ductility and are relatively easy to work with. Carbon percentage for these steels ranges from 0.02 to 0.2. Due to the low carbon content, hardenability cannot be increased by heat treatment and can only be moderately hardened by cold working. Cold work also strengthens the material and reduces its ductility. In comparison these steels in the annealed condition have about 1.5 times more strength than that of carbon steel. In annealed condition they have the best resistance to corrosion. Prolonged exposure of these steels to the temperature range of 400 to 510°C (750 to 950°F) can cause brittleness and loss of notch-impact strength.

Properties of Ferritic Stainless Steel

In annealed state the yield strength of typical ferritic stainless steel is in the range of 240 to 380 MPa (35 to 55 ksi) specially alloy grades like Type 444 and UNS S 44660, and S 44627. These specially alloyed steels, also called superferritic stainless steels, have good resistance to corrosion and are especially good for resistance to stress corrosion cracking (SCC) and pitting.

Pitting Resistance Equivalent (PRE)

Most of these alloys are developed to serve in a specific type of environment and resistance to corrosion is one of them. The resistance to general corrosion of these steels is addressed by alloying elements that develop passivating layers of metallic oxide. However, the localized corrosion is not easy to predict, hence not easy to control. The ability to assess the resistance to pitting of an alloy is determined by a calculated value called Pitting Resistance Equivalent number (PRE).

These alloys exhibit some degree of pitting resistance, which is calculated by a weighted average of key elements like chromium molybdenum, and nitrogen as given in the following equation. A higher PRE number is an indicator of better resistance to pitting and such alloys are preferred where localized corrosion in the form of pitting may be of specific concern:

$$PRE = \%Cr + 3.3(\%Mo) + 16(\%N)$$

It may be noted that the elements include nitrogen, which is a gas, as it has significant influence on the pitting resistance property of an alloy.

Austenitic Stainless Steels

We have discussed basics of stainless steel in the chapter dealing with production of steel. This group of steels is by far the largest group of stainless steel. These steels contain chromium (Cr) in the range of 16 to 26%, nickel (Ni) in the range of 3.55 (grade 201) to 37% (grade 330), other alloying elements including carbon ranging from 0.03% to 0.25%, molybdenum, niobium, titanium, and tantalum, which are specialty additions to improve specific properties like stabilizing or for marine services etc. Other regular alloying elements include manganese, silicon, sulfur, and phosphorus.

Stainless steels are nonmagnetic and can be hardened by cold working. However they cannot be hardened by heat treatment. They have excellent low temperature ductility. Some grades like grade 310 are equally good for moderate to high temperature services.

The chromium-nickel (Type 3xx) and chromium-nickel-manganese (Type 2xx) stainless steels are the two major groups, and they can be hot-worked and cold-worked but they do readily work harden. Austenitic stainless steels are extremely shock-resistant and are not easy to machine.

They exhibit the best high-temperature strength, scaling resistance and superior corrosion resistance compared to the ferritic and martensitic stainless steels. Type 302 is a basic alloy that has been modified to more than 20 basic alloys.

Properties of Austenitic Stainless Steel

The strength of austenitic stainless steel is very much comparable to mild carbon steel, typical yield strength is between 200 to 275 MPa (30 to 40 ksi), the elongation is measured in the range of 40 to 60%. Grade 200 series of steel have better yield strength in annealed conditions 345 MPa to 482 MPa (about 50 to 70 ksi).

The corrosion resistance of these steels is excellent in most of the environment. Of the 3xx series stainless steel the basic grade 304 is commonly used in food, dairy and beverage industries. They are also good in oxidizing environments; the stabilized form of basic, Grade-304, renumbered as Grade 321 (titanium stabilized) and 347 (niobium stabilized) can resist thermally induced embrittlement.

Grade 310 with a higher chromium and carbon level is suitable for higher temperature services; similarly Grade 316 is alloyed with molybdenum to impart resistance from marine environment.

Duplex Stainless Steels

The FeCrNi system alloys have two phases, they typically contain 20 to 30% chromium and 5 to 8% nickel. The alloy system is called Duplex because it typically contains approximately 50/50 austenitic and ferritic phase in their microstructure. These steels contain low carbon and molybdenum, tungsten, copper and nitrogen are added as alloying elements.

Properties of Duplex Stainless Steel

The yield strength of these steels ranges from 550 to 690 MPa (80 to 100 ksi).

The alloy due to its ferritic structure is susceptible to loss of mechanical strength and leads to embrittlement at elevated temperature. Prolonged service at and above 300°C (570°F) is not advised.

The greatest advantage of Duplex stainless steel is the resistance to corrosion, especially to pitting corrosion. The alloys exhibit pitting resistance equivalent (PRE) from 22 (UNSS32304) to 47 (UNS S 32750).

Precipitation-Hardening (PH) Stainless Steels

Precipitation hardening (PH) is defined as caused by the precipitation of a constituent from a supersaturated solid solution. Artificial aging of metal is carried out in which a constituent precipitates from the supersaturated solid solution. These chromium nickel grades of steel are hardened by an aging treatment.

These grades were developed during World War II. They are usually supplied in the solution-annealed condition and are aged for strength following forming. The main grades are 17-4PH or 17-7PH, and the classification is based on their solution annealed microstructure. In annealed condition these grades are semi-austenitic steels, on heat treatment the microstructure is changed to martensite. Cold work enhances the aging effect, certain alloying elements like aluminum, niobium, titanium or copper are added to facilitate aging. ASTM A 286 is the standard to find various grades of precipitation-hardened steels. Some alloys in this class are listed below with their UNS numbers.

UNS Number	Alloy Type
S13800	PH13-8Mo
S15500	15-5 PH
S17400	17-4 PH
S45500	Custom 455

The 17-4PH is solution treated and air cooled to allow the austenite to martensite transformation. Re-heating is performed to increase the strength and corrosion resistance.

Properties of Precipitation-Hardening (PH) Stainless Steel

PH alloys can achieve high strength, ranging from 690 to 1 700 MPa (100 ksi to 250 ksi).

Grade PH 15-7 Mo(h) (UNS S15700) has yield strength of 1590 MPa (230 ksi) and corresponding hardness of above 46 HRC but the elongation which is a measure of ductility, ranges from 1 to 15%, seriously low in all PH alloys. The corrosion resistance of these alloys is, at best, moderate.

Non-Ferrous Materials

COPPER AND COPPER ALLOYS

Copper is used for its high electrical and thermal conductivity, good corrosion resistance, machinability, strength, and ease of fabrication. It can be welded, brazed, or soldered. Copper used for electrical conductors contains over 99.9% Cu and is identified as either electrolytic tough-pitch (ETP) copper or oxygen-free high-conductivity (OFHC) copper. Most copper alloys are homogeneous, single-phase alloys and are not susceptible to heat treatment. Strength is increased by alloying or cold-working.

Brasses are alloys of copper and zinc. The composition may be further alloyed with lead, tin, or aluminum. Alpha (α) brass contains up to 36% Zn and has good corrosion resistance. Yellow α brasses contain 20 to 36% Zn and have high ductility and strength. They are susceptible to dezincification and stress corrosion cracking. Red brasses contain 5 to 20% Zn and have better corrosion resistance than yellow brasses.

Alpha(α) plus Beta(β) brasses contain 54 to 62% copper and have two phases. Muntz metal is the most widely used because it has high strength and good hot-working properties.

Bronze refers to any copper alloy without zinc. Copper alloyed with tin, silicon, aluminum, and beryllium are common bronze alloys.

Cupronickel alloys contain up to 30% nickel and are always single-phase alloys. They have high corrosion fatigue resistance. These alloys also have excellent resistance to seawater corrosion.

ALUMINUM AND ALUMINUM ALLOYS

Aluminum is a non-ferromagnetic, non-pyrophoric and non-toxic material. Its density is 2.7g/cm^3, which is about one-third that of steel. Many aluminum alloys have better strength-to-weight ratios. Aluminum has good malleability and formability, high corrosion resistance, high electrical and thermal conductivity, and is non-sparking. Pure aluminum's tensile strength is about 90 MPa (13,000 psi) and it can be alloyed and heat-treated to increase the strength up to 690 MPa (100,000 psi). These advantages properties have made aluminum an important engineering material. Its use has significantly changed several of the engineering machines and equipment that we use and depend on. Aluminum is not naturally available; in nature it is found in its oxide form (Al_2O_3) and is commonly called alumina. The extraction process is briefly described below.

Aluminum is extracted by a smelting process from bauxite ore by the Hall-Heroult process which involves electrolyzing a bath to extract alumina. The bauxite is dissolved in a cryolite bath, with fluoride salts added to control the temperature. As the electrical current is passed through the bath, the dissolved alumina is electrolyzed. Oxygen is formed which reacts with the carbon anode, and aluminum as the metal collects at the cathode. The metal is periodically siphoned out to crucibles and subsequently cast into ingots. Aluminum recovered by this method is often referred to as primary aluminum. The term secondary aluminum refers to aluminum which has been recovered from scrap. The ingots are further processed and refined to meet industrial specifications. Plates, sheets, foils and extruded shapes and tubs are some of the end products.

PHYSICAL METALLURGY OF ALUMINUM

The physics of aluminum is defined by heat treatment, work hardening, and the effect of alloying elements. Heat treatment and work hardening are the two primary methods used to increase the strength of pure aluminum to make it an engineering material with strength:

- Non-heat treatable alloys are treated by dispersing second-phase constituents in solid solution, and then cold working the alloy
- Heat treatable alloys are treated by dispersing the alloying elements into solid solution, and then precipitating them as coherent submicroscopic particles.

These can be further studied in detail by reviewing the various phase diagrams of different alloys.

EFFECT OF ALLOYING ELEMENTS ON ALUMINUM

Aluminum alloys contain iron silicon and two or more other elements to enhance its properties. The phase formed and the functions obtained are described below.

Effect of Iron

Iron is present in all aluminum as an impurity, generally as a leftover from the smelting process of bauxite. The maximum solubility of iron in aluminum is 0.05%. During the solidification of an iron-aluminum alloy, most of the iron remains in the liquid phase until a eutectic freezes. This consists of solid solution plus Al_3Fe intermetallic constituent particles with a monoclinic crystal structure. Depending on the presence of other alloying elements, e.g., manganese, constituent particles of the metastable orthorhombic Al_6Fe phase can form instead of the Al_3Fe.

Effect of Silicon

Silicon is also present as an impurity. Two ternary phases, cubic $\alpha Al_{12}Fe_3Si$ and monoclinic $\beta Al_9Fe_2Si_2$, are formed by eutectic reaction. At low silicon content, almost all the iron is present in the Al_3Fe phase. As the silicon level is increased, the αAl-Fe-Si phase appears, followed by the βAl-Fe-Si phase.

In larger amounts, alloyed silicon improves the castability and fluidity of the alloy. It is a preferred alloy for brazing sheets. The casting alloys contain silicon in the range of 5 to 20%. Silicon along with manganese allows for precipitation hardening – this is the basis of the 6xxx type of aluminum alloy.

Effect of Manganese

Manganese imparts excellent formability to the alloy. During solidification, some of the manganese (Mn) forms $Al_6(Mn, Fe)$ and cubic $Al_{12}(Mn,Fe)Si$ by eutectic reaction. The remaining Mn remains in solution and is precipitated during the ingot preheat as $Al_{12}(Mn,Fe)Si$ and $Al_6(Mn, Fe)$ dispersoid. These dispersoids strengthen the alloy and control the recrystallized grain size. The Al-Cu-Mn alloy precipitates as $Al_{20}Cu_2Mn_3$ dispersoid particles. The effect of these on the strength is minimal but they are helpful in controlling grain size during the solution heat treatment. The 3xxx types of alloys are an example of Al-Mn alloys.

Effect of Magnesium

The phase diagram of the aluminum magnesium alloy system (Al-Mg), indicates a positively sloping solvus, which is a necessary condition for a precipitation hardening system. The difficulty associated with the nucleation of face centered cubic (fcc) Al_3Mg_2 precipitates has prevented the commercialization of a heat treatable Al + Mg alloy.

Class 5xxx wrought and cast alloys are based on this system. They have excellent strength and corrosion resistance, achieved by solid solution strengthening and work hardening.

Effect of Copper

During solidification some copper combines chemically with aluminum and iron to form either tetragonal Al_7Cu_2Fe or orthorhombic $\alpha(Al.Cu,Fe)$ constituent particles. The subsequent heat treatment cannot dissolve these phases but can transform them from one to another.

The aluminum and copper alloys that contain magnesium (Mg) form an Al_2CuMg phase by eutectic decomposition. Metastable precursors to face-centered orthorhombic Al_2CuMg precipitate-strengthen the alloy. Wrought and cast 2xxx alloys result from this phase; these alloys are desired by the aerospace industry for their strength, fracture toughness and resistance to crack growth.

Effect of Zinc

Aluminum and zinc do not offer much strengthening to alloy, but aluminum, zinc and magnesium precipitates provide two phases that give them strength.

Depending on the zinc:magnesium ratio, the two phases, hexagonal $MgZn_2$ and bcc $Al_2Mg_3Zn_3$ can form by eutectic decomposition in Al-Zn-Mg alloys. The copper-free alloys are strengthened by a metastable precursor to either $MgZn_2$ or $Al_2Mg_3Zn_3$.

In alloys with copper, copper and aluminum replace Zn in $MgZn_2$ to form $Mg(Zn,Cu,Al)_2$. Al_2CuMg particles can also form in these alloys by eutectic decomposition of solid-state precipitation.

Wrought and cast alloys of the 7xxx group are a result of this alloy system.

Effect of Chromium

The solubility of chromium can be reduced to such an extent that primary particles (Al_7Cr) can form by peritectic reaction. These primary particles are harmful to the ductility fracture toughness and fatigue strength of the material, and so the acceptable upper limit of chromium is dependent on the levels of other elements in the system. Chromium dispersoids contribute to strength in the non-heat treatable alloy (of the 5xxx system) that has an fcc structure in $Al_{18}Mg_3Cr_2$ dispersoids. However the alloys that can be heat-treated (e.g. 7xxx) have a dispersoid composition of $Al_{12}Mg_2Cr$.

Effect of Zirconium

This element forms a peritectic with aluminum. The equilibrium Al_3Zr phase is tetragonal, but fine dispersoids of metastable cubic form during preheating treatment of the ingot. Most of the 5xxx, 6xxx and 7xxx alloys have some amounts of zirconium (Zr), usually less than 0.15%, to form an Al_3Zr dispersoid for recrystallization control.

Effect of Lithium

This costly element is used mainly for alloys used in the space industry. It reduces the density and increases the modulus of aluminum alloys. In a binary alloy it forms an Al_3Li precipitate and combines with aluminum and copper in an Al-Cu-Li alloy to form a large number of Al-Cu-Al phases.

AGE HARDENABLE ALLOYS

The following is a list of age-hardenable aluminum alloys, containing various combinations of alloying elements:

Wrought alloys
Aluminum and copper (Al-Cu)
Aluminum, copper and magnesium (Al-Cu-Mg)
Aluminum magnesium and silicon (Al-Mg-Si)
Aluminum, zinc and magnesium (Al-Zn-Mg)
Aluminum, zinc, magnesium and copper (Al-Zn-Mg-Cu).

Casting alloys such as:

Aluminum and silicon (Al-Si)
Aluminum, silicon and copper (Al-Si-Cu).

Work hardenable alloys such as:

Aluminum and magnesium (Al-Mg)
Aluminum and manganese (Al-Mn).

To impart additional properties to aluminum and its alloys, several heat treatment and work hardening processes, or combinations of both, are used.

The temper designations that are used to identify the heat treatment condition for aluminum alloys are listed below:

F = As fabricated
O = Annealed and recrystallized
H = Strain-hardened
W = Solution heat-treated
T = Thermally treated.

Aluminum alloy designations indicate the alloy group, modifications, and heat treatment, where applicable. Unalloyed pure aluminum is classified as 1xxx, and is primarily used in the electrical and chemical industries.

Other aluminum alloy groups are listed below:

1. **Aluminum-Copper Alloys (2xxx Series)** are age hardenable, and include some of the highest strength aluminum alloys, such as Alloy 2024. With a yield strength as high as 66 ksi (455 MPa) its engineering importance is made use of by the aircraft industry.

2. **Aluminum-Manganese Alloys (3xxx Series)** are not heat treatable, have good formability, good corrosion resistance, and good weldability. They are useful for architectural and general purpose applications.

3. **Aluminum-Silicon Alloys (4xxx Series)** are not heat treatable, have excellent castability and corrosion resistance. They are used primarily for making welding and brazing consumables.

4. **Aluminum-Magnesium Alloys (5xxx Series)** are not heat treatable and have good weldability, corrosion, resistance, and moderate strength. Their good corrosion resistance to marine environments makes them a useful material for boat hulls and other applications.

5. **Aluminum-Silicon-Magnesium Alloys (6xxx Series)** are artificially aged and have excellent corrosion resistance and workability. Alloy 6061 is commonly used for structural applications.

6. **Aluminum-Zinc Alloys (7xxx Series)** develop the highest tensile strengths but are susceptible to stress corrosion cracking. Yield strengths exceeding 500 MPa (73 ksi) are achievable; this material is used for high strength applications including aircraft structures.

Aluminum is available in cast and wrought form for various applications. The principal alloying element determines the wrought and cast aluminum designation system as discussed above. They are also designated based on the temper, for both cast and wrought forms. The basic designations are similar to those discussed above. The designations are repeated below:

F As Fabricated
O Annealed
H Strained Hardened
W Solution Annealed
T Solution Heat-Treated

The system designation 'T' is further expanded to include different levels of heat treatment; from T1 to T6.

The strain hardened designation is also further divided into the following grades:

H1 Strained Hardened only
H2 Strained Hardened and Partially Annealed

H3 Strained Hardened and Stabilized
H4 Strained Hardened and Lacquered or Painted

NICKEL AND NICKEL ALLOYS

Nickel is noted for good corrosion resistance, and is particularly useful in oxidizing environments. It forms tough, ductile solid-solution alloys with many metals. Its mechanical properties are similar to those of mild steel, as it retains strength at elevated temperatures while also maintaining ductility and toughness at low temperatures.

Nickel-Copper alloys such as Cupro-Nickel have excellent corrosion resistance in a wide variety of environments and may be used at temperatures up to 815°C (1,500°F).

Nickel-Silicon alloys such as Hastelloy-D are strong, tough, and extremely hard. They have excellent corrosion resistance in sulfuric acid at high temperatures.

Nickel-Chromium-Iron alloys combine the corrosion resistance, strength, and toughness of nickel with the high temperature oxidation resistance of chromium.

Nickel-Molybdenum-Iron alloys such as Hastelloy-B have high corrosion resistance to hydrochloric, phosphoric, and other non-oxidizing acids.

Nickel-Chromium-Molybdenum alloys such as Hastelloy-C have high corrosion resistance to oxidizing acids, good high temperature properties, and are resistant to oxidizing and reducing atmospheres up to 1,093°C (2,000°F).

TITANIUM AND TITANIUM ALLOYS

The density of Ti is about 0.16 lb/in^3 compared to steel's 0.28 lb/in^3. This highlights the excellent strength-to-weight ratios of titanium alloys. Titanium has excellent corrosion resistance up to 538°C (1,000°F). Alloying elements influence the alpha-to-beta transition temperature and so are referred to as either alpha or beta stabilizers. Single-phase alloys are weldable and have good ductility. Some two-phase alloys are also weldable but experience loss of ductility. Two-phase alloys can be strengthened by heat treatment.

Working with Metals

Working with metals involves understanding the limits of their mechanical properties. Several aspects of these properties can be used to get the best results from the metal. There cannot be a limit on what and how much information to know about any given material. As they can be used in very different environmental conditions, the impact of the environment on the metal must be studied in order to determine which properties of the material would be affected.

A combination of information, including a material's mechanical properties and corrosion behavior in the specific service environment must be studied for the proper selection of a material for any project. Some of the mechanical properties that have a significant impact on the workability of metals are discussed in this chapter.

ELASTIC LIMIT

When a material is stressed below its elastic limit, the resulting deformation or strain is temporary. Removal of an elastic stress allows the object to return to its original dimensions. When a material is stressed beyond its elastic limit, plastic or permanent deformation takes place and it will not return to its original dimensions when the stress is removed. All shaping operations, such as stamping, pressing, spinning, rolling, forging, drawing, and extruding involve plastic deformation. Pressure testing, with few exceptions, is done within the elastic limits of the material.

PLASTIC DEFORMATION

Plastic deformation may occur by slip, twinning, or a combination of the two. Slip occurs when a crystal is stressed in tension beyond its elastic limit. It elongates slightly and a step appears on the surface, indicating the displacement of one part of the crystal. Increasing the load will cause movement on a parallel plane, resulting in another step. Each successive elongation requires a higher stress and results in the appearance of another step. Progressively increasing the load eventually causes the material to fracture.

Twinning is a movement of planes of atoms such that the lattice becomes divided into two symmetrical parts that are differently oriented. Deformation twins are most prevalent in close-packed hexagonal metals, such as magnesium and zinc, and body-centered cubic metals, such as tungsten and α iron. Annealing twins can occur as a result of reheating previously worked face-centered cubic metals such as aluminum and copper.

FRACTURE

Fracture is the separation of a body under stress into two or more parts. Brittle fracture involves rapid propagation of a crack with minimal energy absorption and plastic deformation. It occurs by cleavage along particular crystallographic planes and shows a granular appearance.

Ductile fracture occurs when considerable plastic deformation takes place prior to failure. Fracture begins by the formation of cavities at nonmetallic inclusions. Under continued applied stress, the cavities coalesce to form a crack. This process is seen as micro-void coalescence on the fracture surface.

POLYCRYSTALLINE MATERIALS

Commercial materials are made up of polycrystalline grains whose crystal axes are oriented at random. Therefore, depending on their orientation to the applied stress, the deformation processes occur differently in the grains.

A fine-grained material in which the grains are randomly oriented will possess identical properties in all directions and is called isotropic.

A metal with controlled grain orientation will have directional properties (anistropic) which may be either troublesome or advantageous, depending on the direction of loading.

When a crystal deforms, there is distortion of the lattice which increases with increasing deformation. As a result, there is an increase in resistance to further deformation known as strain hardening or work hardening.

COLD WORKING

A material is considered to be cold worked when its grains are in a distorted condition after plastic deformation has occurred. All of the properties of a metal that are dependent on the lattice structure are affected by plastic deformation.

Cold working increases the tensile strength, yield strength, and hardness of the material. Hardness increases most rapidly in the first 10% reduction by cold work and tensile strength increases linearly; conversely most of the ductility is lost due to cold work in the first 10% reduction; thereafter the reduction in ductility happens more slowly. Yield strength increases more rapidly than tensile strength. Cold work also reduces electrical conductivity.

STORED ENERGY

Although most of the energy used to cold-work metal is dissipated in heat, a finite amount is stored in the crystal structure as internal energy that is associated with the lattice defects created by the deformation. This increase in internal energy is often concentrated in the grain boundaries, resulting in localized increased susceptibility to energy-driven reactions such as corrosion.

RESTORING THE LATTICE STRUCTURE OF METAL AFTER COLD WORK – ANNEALING

Full annealing is the process by which the distorted, cold-worked, lattice structure is changed back to one which is strain-free through the application of heat. This is a solid-state process and is usually followed by slow cooling in the furnace.

Recovery is the first stage of annealing. This is a low temperature process and does not involve significant changes in the microstructure. The principal effect is relief of internal stresses. Recovery is a time and temperature dependent process. There is little change in mechanical properties and the principal application of recovery is stress relief to prevent stress corrosion cracking or to minimize distortion produced by residual stresses.

Recrystallization occurs at higher temperatures, as minute new crystals appear in the microstructure. They usually appear in the regions of highest deformation such as at grain boundaries or slip planes. Recrystallization takes place by the process of nucleation of strain-free grains and the growth of these nuclei to absorb the cold-worked material.

The recrystallization temperature refers to the approximate temperature at which a highly cold-worked material completely recrystallizes in one hour. It may be noted that the greater the amount of deformation, the lower the recrystallization temperature. Zinc, lead, and tin have recrystallization temperatures below room temperature and so cannot be cold-worked.

GRAIN GROWTH

Large grains have lower free energy than small grains, since there is less grain boundary volume. Grain growth is driven by a single crystal's lowest energy state. The rigidity of the lattice opposes grain growth.

As temperature is increased, the rigidity of the lattice decreases and the rate of grain growth is more rapid. Holding a specimen for a long time in the grain-growth temperature region (slightly below the melting point) can grow very large grains. The final recrystallized grain size is controlled by factors that influence nucleation and growth rate.

HOT WORKING

When a material is plastically deformed, it tends to become harder. But the rate of work hardening decreases as the working temperature is increased. Two opposing effects take place – hardening due to plastic deformation and softening due to recrystallization.

For a given material, there is a temperature at which these two effects balance. Material worked above this temperature is said to be hot-worked and material that is worked below this temperature is said to be cold-worked. Lead and tin may be hot-worked at room temperature, whereas steel is cold-worked at 538°C (1,000°F).

Hot-worked material cannot be manufactured to an exact size because of the dimensional changes which take place during cooling. Cold-worked materials can be held to close tolerances, but require more power for deformation and so are more expensive to produce.

Normally, initial reductions are carried out at an elevated temperature by hot working and the final reductions are done cold to take advantage of both processes. The finishing temperature for hot working determines the grain size that is available for further cold working. Careful control of these processes is known as thermo-mechanical processing.

Mechanical Properties and Testing of Metals

The science of the behavior of metals and alloys subjected to applied forces is known as mechanical metallurgy. One approach uses the strength of the material and the theories of elasticity and plasticity, and considers a metal to be a homogeneous material whose mechanical behavior can be described on the basis of only a few material constants. However, the theories of strength of materials, elasticity and plasticity lose much of their power when the material's structure becomes important and it can no longer be considered as a homogeneous medium.

STRENGTH OF MATERIALS

With respect to metallurgy, knowledge of the strength of a material comes from the relationship between internal forces, deformation, and external loads.

The material is assumed to be in equilibrium, and the equations of static equilibrium are applied to the forces acting on some part of the body in order to obtain a relationship between the external forces and the internal forces resisting their action.

Although metals are made up of an aggregate of crystal grains having different properties in different directions, the equations of strength do apply, since the crystal grains are so small that the materials are statistically homogeneous and isotropic on a macroscopic scale.

ELASTIC AND PLASTIC BEHAVIOR

All solid materials can be deformed when subjected to external load. In the previous chapter we discussed the material's elastic and plastic deformation and its impact on the formability of metals.

The recovery of the original dimensions of a deformed body when the load is removed is called elastic behavior.

The limiting load beyond which material no longer behaves elastically is the elastic limit.

A body which is permanently deformed is said to have undergone plastic deformation.

For most materials which are loaded below the elastic limit, the deformation is proportional to the load in accordance with Hooke's Law, which in simple terms says that strain is directly proportional to stress.

Mathematically, this law can be expressed as:

$$F = -kx,$$

Where:

x is the displacement of the end of the spring from its equilibrium position (meters [m] in SI units);

F is the restoring force exerted by the material (Newtons [N] or $kgms^{-2}$ or kgm/s^2 in SI units);

k is a constant called the *rate* or *spring constant* ($N·m^{-1}$ or kgs^{-2} or kg/s^2 in SI units).

DUCTILE VS. BRITTLE BEHAVIOR

A completely brittle metal would fracture almost at the elastic limit and a mostly brittle material such as white cast iron would show some measure of plasticity before fracturing.

Adequate ductility is an important consideration, since it allows the material to redistribute localized stresses. If localized stresses at notches and other stress concentrations do not have to be considered, it is possible to design for static situations on the basis of average stress.

With brittle materials, localized stresses continue to build up when there is no local yielding. Finally, a crack forms at one or more points of stress concentration and spreads rapidly over the section.

Even without a stress concentration, fracture occurs rapidly in a brittle material, since the yield stress and tensile strength are nearly the same.

FAILURE

Structural members fail in three general ways:

- Excessive elastic deformation
- Yielding or excessive plastic deformation
- Fracture.

Failure due to excessive elastic deformation is determined by the modulus of elasticity rather than the strength of the material.

Yielding, or excessive plastic deformation, occurs when the elastic limit is exceeded. Yielding rarely results in fracture of a ductile metal, since the metal strain-hardens as it deforms and an increased stress is required to produce further deformation.

At elevated temperatures, metals no longer exhibit strain hardening and can continuously deform at constant stress – known as creep.

Metals fail by fracture in three general ways:

- Sudden brittle fracture
- Fatigue fracture or progressive fracture
- Delayed fracture.

A change from ductile to brittle behavior can occur when the temperature is decreased, the rate of loading is increased, and a notch forms a complex state of stress.

Most fractures in machine parts are due to fatigue. A minute crack starts at a localized spot (notch or stress concentration) and gradually spreads over the section until it breaks. There is no visible sign of yielding at the average stresses, which are often below the tensile strength of the material.

Delayed fracture can occur as stress-rupture in a statically loaded material at elevated temperature over a long period of time. Static loading in the presence of hydrogen can also cause delayed fracture.

FRACTURE

Fracture is the separation of a solid body into two or more parts under the action of stress. The fracture process has two components: crack initiation and crack propagation.

Ductile fracture is characterized by appreciable plastic deformation prior to and during the crack propagation.

Brittle fracture is characterized by a rapid rate of crack propagation with no gross deformation and very little micro-deformation.

For steel, the possibility of brittle fracture increases with decreasing temperature, increasing strain rate, and triaxial stress conditions usually produced by a notch.

Different terms are used to characterize fractures:

Behavior Described	Terms Used	
Crystallographic mode	Shear	Cleavage
Appearance	Fibrous	Granular
Strain to fracture	Ductile	Brittle

FRACTURE CONTROL

Fracture control is a combination of measures to prevent fracture due to cracks during operation. It includes damage tolerance analysis, material selection, design improvement, and maintenance and inspection schedules.

Damage tolerance analysis has two objectives:

- To determine the effect of cracks on strength
- To determine the crack growth as a function of time.

The effect of crack size on strength is denoted as a length, and strength is expressed in terms of the load, P, that the structure can carry before fracture occurs.

A new structure with no defects ($a = 0$) will have a strength P_u (the ultimate design strength or load). If the maximum anticipated service load is P_s, a safety factor j is applied so that the structure can actually sustain a load jP_s.

If cracks are present, the strength of the material is less than P_u. This remaining strength is called the residual strength P_{res} and the diagram is called a residual strength diagram.

With a crack of size a, the residual strength is P_{res} and a load greater than P_{res} will cause fracture to take place. The fracture process may be slow and stable at first and the structure can hang together. Eventually the fracture becomes unstable and the structure breaks. The change from stable to unstable fracture may occur in a few seconds.

At loads below P_{res} the crack will continue to grow by other means than fracture such as fatigue, stress-corrosion, or creep.

As crack growth continues, the crack becomes longer, the residual strength lower, the safety factor lower, and the probability of fracture higher.

The crack must not be allowed to grow so large that fracture occurs at the service load, or even at P_a. The maximum permissible crack size follows in terms of the calculated residual strength and the prescribed minimum permissible strength P_p.

The residual strength diagram differs for different components of the structure and for different crack locations.

A critical crack a_c is one that will cause fracture in service. The value a_p is the tolerable crack size that will not allow fracture at the design load.

Crack growth curves are used to determine when a crack might reach a_p. At some crack size a_o, the crack grows in time, and use of the crack growth curve allows one to predict when a_p will be reached.

CRACK GROWTH AND FRACTURE

Crack growth occurs slowly during normal service loading. Fracture is the final event and often takes place very rapidly.

Crack growth takes place by one of five mechanisms:

- Fatigue, due to cyclic loading
- Stress corrosion due to sustained loading
- Creep
- Hydrogen induced cracking
- Liquid metal induced cracking.

Even at very low loads there is still plastic deformation at the crack tip because of the high stress concentration.

Crack growth by stress corrosion is a slow process in which the crack extends due to corrosive action (often along grain boundaries) facilitated by atomic disarray at the crack tip.

Fracture can only occur by one of two mechanisms; cleavage or rupture. Cleavage is the splitting apart of atomic planes. Each grain has a preferred plane and the resultant fracture is faceted.

Ductile rupture occurs around particles and inclusions, which are always present in a metal or alloy. First, the large particles become loose or break away from the bonding, resulting in widely spaced holes close to the crack tip.

Eventually, voids are formed at many locations and these voids eventually join up to form a fracture. The resultant fracture surface looks dull.

Both cleavage and rupture are fast processes. Cleavage fractures can travel as fast as 1 mile/sec and dimple ruptures as fast as 1,500 ft/sec.

Damage Tolerance

Fracture mechanical methods have been developed to analyze fracture, to obtain the fracture stress (residual stress), and to analyze fatigue crack growth and stress corrosion crack growth. Creep crack growth is still in the research phase.

In stress corrosion cracking, crack growth times are usually so short that crack growth analysis is often unimportant and the focus is on crack growth prevention.

FAILURE ANALYSIS

Failure analysis is a forensic analysis of failed parts. It has various possible objectives including academic interest, to develop a better material, to assign responsibility for the cause of the failure, or to meet mandatory or regulatory compliance etc. Failures, as discussed above, can cause damage to life and property and it is sometimes necessary to conduct failure analysis as a legal necessity.

The process of failure can vary from simple to complex, and hence its analysis may involve coordination of experts from several disciplines. This may include various disciplines of engineering and may well extend to other sciences including, for example, mathematical modeling, physics or chemistry.

The process involves planning a failure investigation, sequencing the activities, drawing up a team and planning and coordinating their activities and progress. The leader of the investigation is required to constantly review progress and the findings of each expert, and if necessary correct the course of the investigation in order to meet the goals of the analysis.

Planning involves the following activities:

1. Reviewing the objective of the failure analysis.
2. Selecting the team and assigning the role of each individual member of the team.
3. Estimating the required tools and equipment and providing those to the members of the team.
4. Identifing the tests, examinations and inspection requirements that would be required for the analysis.

After the plan for the failure analysis is made, it is time to start implementing it. The plan needs to be flexible to accommodate any new information that may emerge during the course of the investigation. A list of activities may be followed, but it must be noted that there is no fixed route for a good failure investigation; each failure has its own personality and will require a very carefully planned activity.

1. Collect of all background information.
2. Investigate events leading to the failure.
3. Collect all data, including engineering drawings, sketches drawn, and pictures taken, and the notes made of conversations with witnesses and field personnel involved with the failure site.
4. Acquire details of the history of the part, including the initial manufacturing process, any repair or any previous failures of the part, or any complementing part that may have a bearing on the current failure.
5. Conduct an inspection of the part and the general layout of the site and machinery involved.
6. Understand the process and environment in which the equipment is put to service.
7. Start the investigation work.
8. Review the resulting data, test reports, and compare the available information.
9. Derive conclusion.
10. Compile the report giving the background information, describing the failure with mode of failure and cause of failure. And finally, recommend corrective and preventive action.

TESTING OF METALS

Tensile Test

Tensile testing is used to establish operational load limits for metals and alloys. A sample of the material is prepared so that a force can be applied along its axis. A central portion of the sample is reduced in width so that it will experience the highest stresses.

The tensile test measures the ability of a material to support a stress (force per unit area). The response of a tensile sample to the application of an increasing stress can be described in terms of elastic and plastic behavior. Initially the sample undergoes elastic elongation as it is pulled. As increasing stress is applied, the sample undergoes permanent deformation; that is plastic strain. A stress-strain curve is used to determine the point at which the reversible elastic strain is exceeded and permanent or plastic deformation occurs. The yield strength is the stress necessary to cause significant plastic deformation (usually defined as 0.2% strain).

Young's modulus, or the modulus of elasticity, states the relationship between stress and strain represented by the straight-line portion of the stress-strain curve. It reflects the tendency of a material to deflect under a given applied stress.

Once yielding has begun, there is significant motion of dislocations within the metal grains. Grain boundaries, phase boundaries, and other dislocations hinder their movement. As more obstacles are encountered, the stress necessary to cause continued dislocation movement becomes greater. The maximum stress that the material can withstand before breaking is the ultimate tensile strength.

Reduction of area and percent elongation can be calculated from the broken sample. They are both indicators of a material's ductility.

Hardness Test

Hardness is defined as the resistance to plastic deformation by indentation. Brinell, Rockwell, and Vickers or Knoop are the most common indentation hardness test methods. The depth or width of the impression left by the indentation is measured to indicate hardness.

The Brinell test presses a very hard ball into the surface under a standard load. Since a large impression is made, it is suitable only for large samples.

The Rockwell test forces a cone or sphere into the surface under a standard load. Different indenters and different scales are used for different hardness ranges.

Knoop and Vickers rests are called micro-hardness tests because of the very small indentation size. They are used to measure hardness on small samples and in small specific areas such as the heat affected zone of a weld.

Rebound hardness tests measure the height to which a ball rebounds after being dropped on a surface. Equotip is a commercial rebound tester.

Impact Test

These tests measure the ability of a material to absorb energy during sudden loading in order to evaluate its tendency to brittle fracture.

A heavy mass is positioned above the sample and allowed to strike the sample upon release. The difference between the potential energy of the mass before and after impact (i.e., the energy absorbed by the impact and fracture) is calculated and is called the impact energy. It is an indicator of the material's toughness.

Charpy and Izod samples are used. Both are notched, so that breaking will occur in the controlled area.

In a bcc material like steel, the impact energy is reduced as temperature is reduced. A series of tests done at progressively lower temperatures can be performed to determine the temperature at which the fracture mode changes from ductile to brittle. Such a point on the chart is called the transition temperature, as it relates to the change in the ductility of the material.

Creep Test

Creep is time-dependent plastic deformation that occurs at stresses below the yield strength of the material, and in steel is normally of significance only at elevated temperatures.

A tensile specimen that is loaded in tension below its yield strength and then heated will elongate with time. The creep strain (or elongation) against the time is plotted to generate a creep curve.

If time-to-failure is the parameter of interest, the test is called a stress-rupture test.

Fatigue Test

Failure under repeated loads is called fatigue failure. Typically, fatigue cracks initiate at some defect in the part and propagate through it as a result of a cyclic stress. Iron-based alloys exhibit a fatigue or endurance limit – a stress below which the part can theoretically be cycled infinitely without failure.

Fatigue in weld is important in determining the performance and life of a structural member. Fatigue testing of weld is discussed in some detail in Chapter 5 of Section 2 in this book.

Heat Treatment of Steels

Heat treatment is the practical application of physical metallurgy discussed earlier in this book. The action is in most cases thermal, with the aim of modifying the structure of the material in order to obtain desired properties. Other approaches could include thermo-chemical and thermo-mechanical processes, both of which strive to reach the same goal as the plain controlled heating and cooling cycle that is generally understood by the term 'heat treatment'.

Heat treatment is defined as a combination of heating and cooling operations, timed and applied to a metal or alloy in the solid state in a way that will produce desirable properties.

The first step in the heat treatment of steel is to heat the material to some temperature at or above the critical range in order to form austenite. Different heat treatments are based on the subsequent cooling and reheating of the austenitized material.

To understand the heat treatment of steel – and for that matter any metal – it is important to fully understand the phase diagram of that metal. We have discussed phase diagrams in detail in Chapter 4 of this section, and now shall take the most common and most applied phase diagram of iron carbon to discuss the heat treatment of steel. Where necessary, other materials will be included to emphasize a point.

As stated above, phase diagrams are important tools for understanding the heat treatment process. Several different diagrams and charts can be derived from the basic all-encompassing diagram; some of which are very specific to a given material, temperature or set of properties. For example, an isothermal (IT) diagram describes the formation of austenite, and is called an ITh diagram, and the one that describes the decomposition of austenite is referred to as a time temperature transformation (TTT) diagram. Continuous cooling and continuous heating diagrams, commonly referenced as CCT and CHT diagrams, are also used to understand the effects of heating and cooling steel during the heat treatment process.

During heat treatment there is the possibility of both relaxation of residual stresses, e.g., stress relieving, discussed further in this chapter, and introducing additional stress caused by thermal expansion and contraction. The thermal strain developed is directly proportional to the cycle of heating and cooling. Such a relationship is called the thermal strain (σ_{th}), and is easy to determine, if we know the change in heat (ΔT), the elastic modulus of the material (E), and the thermal coefficient of expansion (α):

$$\sigma_{th} = \alpha \Delta TE$$

It may be noted that if the thermal stress exceeds the flow strength of cooler or hotter material, then permanent plastic deformation may occur. Apart from the dimensional problems, the possible phase changes may induce fundamentally different behaviors and properties in the resulting material.

TTT AND CCT CURVES

Isothermal-Transformation (ITh) or (TTT) Diagrams

As is discussed above, the phase diagram is of little use for steels that have been cooled under non-equilibrium conditions; isothermal transformation (ITh) or time temperature transformation (TTT) diagrams have been developed to predict non-equilibrium structures. The TTT diagram of eutectoid plain carbon steel is described below.

The curve is prepared by heating steel samples to the austenitizing temperature of 1,550°F (845°C). These samples are then quenched to a set of temperatures below 1,300°F (700°C). A molten bath is used in the laboratory for the control of temperature.

In the above two TTT diagrams for steel we can see that above the A_e, austenite is stable. The area to the left of the beginning of the transformation consists of unstable austenite. The area to the right of the

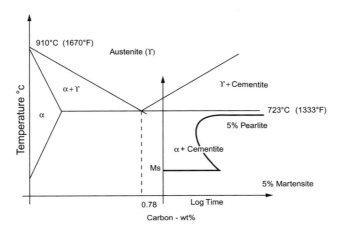

FIGURE 1-12-1   TTT curve of eutectoid steel superimposed on iron carbon phase diagram.

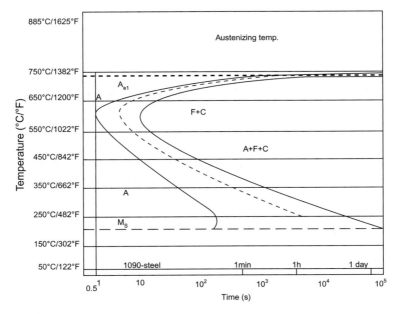

FIGURE 1-12-2   Time temperature transformation diagram for 1090 steel. *Reprinted with permission of ASM International. All rights reserved.*

end-of-transformation line is the product to which austenite will transform at constant temperature.

The area between the beginning and the end of the transformation, labeled A + F + C, consists of three phases: austenite, ferrite, and carbide. The Ms temperature is indicated as a horizontal line, and temperatures for 50% and 90% transformation from austenite to martensite are indicated.

The transformation product above the nose region is pearlite. As the transformation temperature decreases, the spacing between the carbide and ferrite layers becomes smaller and the hardness increases. Between the nose region of 510°C (950°F) and the Ms temperature, an aggregate of ferrite and cementite appears which is called bainite. As the transformation temperature decreases, the bainite structure becomes finer.

COOLING CURVES

In heat treatment, the control of cooling is as important as the heating. The processes require a well-established control program to achieve the desired properties of the metal. Constant cooling rate diagrams have been developed. Curves for natural cooling, based on Newton's law of cooling, are also developed and used. These simulate the cooling rate Jominy bar behavior.

Various cooling curves can be superimposed on the I-T diagram. In the following, a set of cooling curves have been superimposed on a TTT diagram. This set is specifically developed for martempering steel.

In the first curve (a) a very slow cooling rate typical of annealing is indicative of conventional process of tempering. The second curve (b) indicates martempering and the third curve (c) is a curve for modified martempering.

COOLING-TRANSFORMATION (C-T) DIAGRAMS

The actual heat treatment of steel involves continuous cooling, and so the C-T diagram has been derived from the I-T diagram.

In general, the nose will be shifted downward and to the right as a result of faster cooling rates. Since C-T diagrams are much more difficult to derive and I-T diagrams normally err conservatively, I-T diagrams are often used to predict structures and to determine critical cooling rates.

Chemical composition and austenite grain size affect the position of the I-T curve. Increasing levels of carbon or alloying elements moves the curve to the right; transformation is retarded and critical cooling rates are slowed down, making martensite easier to obtain.

STRESS RELIEF ANNEALING

As the name suggests the treatment is given to steel to relive the locked-in stresses due to the manufacturing processes of forging, rolling, bending, stamping, forming, welding etc.

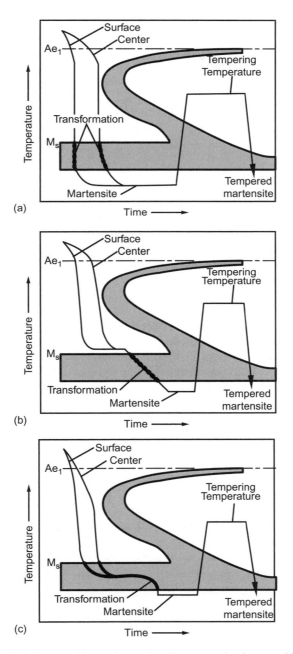

FIGURE 1-12-3   TTT diagram with superimposed cooling curves showing quenching and tempering (a) Conventional process (b) Martempering (c) Modified.   *Reprinted with permission of ASM International. All rights reserved.*

The process involves heating the affected metal – either entire or a partial section – to a temperature that is below the transformation range. It is usually carried out at temperatures below the lower critical line (AC$_1$) 723°C (1,333°F); the typical carbon steel stress relief temperature is 600°C to 675°C (1,000°F to 1,250°F). The steel is then held for sufficient time to soak the entire thickness to a predetermined temperature, and then cooled uniformly throughout the cross section and surface area. As is evident, the stress relief is a time/temperature phenomenon. The process may involve either microscopic or even macroscopic creep relaxation at the stress relief temperature.

NORMALIZING

The objective of normalizing varies. It can increase or decrease the strength and hardness of steel, homogenize and refine the grains, or reduce residual stresses. The effect is dependent on the history of the material, and the heating and cooling cycle practiced. It is often seen as overlapping the function of various other types of heat treatment, such as annealing, hardening and stress relieving.

The process involves austenitizing the steel, followed by controlled heating under still air. The process is both a thermal and microstructural phenomenon. Normalizing is carried out at about 55°C (100°F) above the upper-critical-temperature (A$_3$ line), followed by cooling in still air.

Normalizing produces a harder and stronger steel, improves machinability, modifies and refines cast dendritic structures, and refines the grain size for improved response to later heat treatment operations. Since cooling is not performed under equilibrium conditions, there are deviations from the structures predicted by the phase diagram.

ANNEALING

The full annealing process consists of heating to the proper temperature and then cooling slowly, through the transformation range, in the furnace. The purpose of annealing is to produce a refined grain, to induce softness, improve electrical and magnetic properties, and sometimes to improve machinability. Annealing is a slow process that approaches equilibrium conditions and comes closest to following the phase diagram.

Annealing involves several thermal cycles, classified on the basis of the maximum temperature reached:

- **Subcritical Annealing:** Heating below the lower critical temperature A$_1$.
- **Intercritical Annealing:** Heating above A$_1$ but below upper critical temperature A$_3$ for hypoeutectoid and A$_{cm}$ for hypereutectoid steels.
- **Full Annealing:** Heating above upper critical temperature A$_3$.

The table below indicates the recommended temperatures and cooling cycle for full annealing of various carbon steel forgings.

TABLE 1-12-1 Full Annealing Temperature Cycle for Forgings

AISI Steel	Annealing Temperature Celsius	Cooling Cycle From Celsius	To	Hardness Range HB
1018	855–900	855	705	111–149
1020	855–900	855	700	111–149
1025	855–900	855	700	111–187
1030	845–885	845	650	126–197
1040	790–870	790	650	137–207
1050	790–870	790	650	156–217
1060	790–845	790	650	156–217
1070	790–845	790	650	167–229
1080	790–845	790	650	167–229
1090	790–830	790	650	167–229

Reprinted with permission of ASM International. All rights reserved.

SPHEROIDIZING

In hypereutectoid steels, the cementite network is hard and brittle and must be broken by the cutting tool during machining. Spheroidizing annealing is performed to produce a spheroidal or globular form of carbide and improve machinability. All spheroidizing treatments involve long times at elevated temperatures, to produce a structure of globular carbide in a ferritic matrix. Application of spheroidizing is discussed in the chapter about cast iron.

In general spheroidizing is accomplished by the following procedures:

1. Heating to a temperature just below Ae_1 and holding at that temperature for a prolonged time.
2. Alternate heating and cooling between temperature range of just above Ac_1 and just below Ar_1.
3. Heating to a temperature above Ac_1, followed by either slow cooling in the furnace or holding at a temperature just above Ar_1.
4. Cooling at a suitable rate from the minimum temperature at which carbide is dissolved to prevent reformation of carbide network, followed by reheating in accordance with either method number one or two above. This method is applicable to steels that have carbide networks, such as hypereutectoid steels.

TEMPERING

Steel in the as-quenched martensitic condition is too brittle for most applications. High residual stresses are induced as a result of the martensite transformation. Therefore, tempering or drawing nearly always follows hardening.

Tempering involves heating the steel to some temperature below the lower critical temperature and thus relieving the residual stresses and improving its ductility and toughness. There is usually some sacrifice of hardness and strength.

Hardness decreases and toughness increases as the tempering temperature is increased. However, toughness drops when the tempering is performed in the temperature range of 200°C to 430°C (400°F to 800°F). Residual stresses are largely relieved during tempering in this temperature range.

Certain alloy steels exhibit temper brittleness when tempered in the 538°C to 675°C (1,000°F to 1,250°F) range followed by slow cooling. Molybdenum has a retarding effect on temper embrittlement.

Martensite is a supersaturated solid solution of carbon trapped in a body-centered tetragonal structure and is metastable. As energy is applied during tempering, the carbon is precipitated, and the carbide and the iron will become bcc. As the tempering temperature is raised, diffusion and coalescence of the carbides occurs and the final product is called tempered martensite.

Different types of tempering are listed below with brief descriptions.

Austempering of Steels

Carbon and low alloy steels are austempered prior to hardening. The important factor to note about the austempering process is that the rate of heating assumes secondary importance to the maximum temperature attained throughout the section of the steel, the holding (soaking) time at that temperature, and the cooling rate. The thermal conductivity of steel, the thickness of the section, spaced or stacked furnace loading method, the nonscaling or scaling atmosphere of the furnace, and the circulation of heat in the furnace are the important variables for the success of this tempering process.

MARTEMPERING

The term martempering is used for a process that is carried out by heating the steel and castings to an elevated temperature (austenitizing) and quenching in a hot fluid medium usually kept above the martensitic M_s point, and stabilizing the material at that temperature by holding it for sufficient time in the medium. Air cooling at slow rate follows, to maintain an easy temperature gradient between outer and inner heat at the cross section of the steel or casting. The process involves formation of martensite, hence avoidance of residual stresses is of great importance. This is followed sequentially by cooling to room temperature, and subsequent conventional tempering, to temper the primarily brittle martensite microstructure. Figure 1-12-3b above describes the difference between conventional quenching and martempering.

The procedure's principal aim is to reduce residual stresses, distortions and the development of cracks.

HARDENING

Hardenability is related to the depth of penetration of the hardness. It is predicted by the Jominy test. The test involves heating a 4-inch long bar uniformly to the proper austenitizing temperature, then quenching by a controlled water spray.

A plot of the hardness vs. distance from the quenched end is made. Since each spot on the test piece represents a certain cooling rate and since the thermal conductivity of all steels is assumed to be the same, the hardness at various distances can be used to compare the hardenability of a range of compositions.

Hardening by Martensite Transformation

Under slow or moderate cooling rates, carbon atoms have time to diffuse out of the (fcc) austenite structure so that the iron atoms can rearrange themselves into the bcc lattice. This γ to α transformation takes place by nucleation and growth, and is time dependent.

Faster cooling rates do not allow sufficient time for the carbon to diffuse out of solution, the structure cannot transform to bcc, and the carbon atoms remain trapped in solution. The resultant structure – martensite – is a supersaturated solid solution of carbon trapped in a body-centered tetragonal structure. This is a highly distorted structure that results in high hardness and strength.

Martensite atoms are less densely packed than austenite atoms, so a volumetric expansion occurs during the transformation. As a result, highly localized stresses produce distortions in the matrix. The transformation is diffusionless, and small volumes of austenite suddenly change crystal structure by shearing actions. It proceeds only during cooling and stops if cooling is interrupted. The temperature of the start of martensite transformation is known as the Ms temperature, and at the end of the transformation, the Mf temperature. Martensite transformation cannot be suppressed or the Ms temperature changed by changing the cooling rate. Ms is a function of chemical composition only.

Martensite is never in a state of equilibrium, although it can persist indefinitely at or near room temperature. It would eventually decompose into ferrite and cementite.

CASE HARDENING AND CARBURIZING

Another hardening method called case hardening involves carburizing the outer skin of the material to attain a hard outer surface (case), while retaining the softer and tougher inner core. The process requires a carbon source and a furnace to attain the required temperature in a controlled environment that may be in some cases the carburizing gases, preparation of the components to be case hardened.

Many steel forgings are heat-treated to develop a hard, wear-resistant surface while retaining the soft tough core. Such properties can be obtained in low carbon steel forgings by case-hardening, using either a carburizing or nitriding process.

The process variables for carburizing include:

1. Accurately controlled temperature to achieve the required degree of hardness. The diffusion rate of carbon in steel increases with increasing temperature. The increase in carbon absorption from 870°C (1,600°F) to 925°C (1,900°F) is about 40%.
2. The time over which the temperature is held to austenitized the case of the steel is an important variable that determines the depth of the hardness in the case. Longer times will austenitize deeper in the case, thus forming a deeper hardened case. The table below shows the effect of time on the depth of thickness of the steel case.

TABLE 1-12-2 Carburizing Case-Depth vs. Time and Temperature

Time (Hours)	870 deg C (mm)	900 deg C (mm)	927 deg C (mm)	955 deg C (mm)
1	0.46	0.53	0.64	0.74
2	0.64	0.76	0.89	1.04
4	0.89	1.07	1.27	1.3
8	1.27	1.52	1.8	2.11
12	1.55	1.85	2.21	2.59
16	1.8	2.13	2.54	2.97
24	2.18	2.62	3.1	3.66
30	2.46	2.95	3.48	4.09

3. Carburizing atmosphere. The main constituents of the atmosphere are CO, N_2, H_2, CO_2 H_2O, CH_4 out of all these only nitrogen (N_2) is inert; the remainder are present in the atmosphere at equilibrium with the reversible reaction.

Various case hardening methods are listed below:

- Gas carburizing. The carbonitriding or gas nitriding process involves the use of ammonia gas (NH_3) to produce nascent nitrogen that enters the case of the steel and increases its hardness. The nitriding temperature ranges from 450°C to 620°C (850 to 1,150°F), but the best results are obtained at about 510°C (950°F). The process of nitriding is suited to alloy steels, whereas the carburizing process is most suited to low-carbon steel.
- Pack carburizing
- Liquid carburizing (cyaniding) using potassium cyanide (KCN) or sodium cyanide (NaCN) in a bath at about 845°C (1,550°F). At this temperature the steel will absorb carbon and nitrogen from the bath to a depth of about 0.125 mm (0.005 inches) within 15 minutes. A longer time in the bath will develop hardness to a depth of about 0.25 mm (0.10 inches). To give full hardness to the case (surface) of the steel, the material is quenched in water or brine immediately after being removed from the cyanide bath.
- Vacuum carburizing also involves a plasma (ion) process.

PROCESS OF QUENCHING

Cooling proceeds through three separate stages during a quenching operation. Vapor-blanket cooling describes the first cooling stage, when the quenching medium is vaporized at the metal surface and cooling is relatively slow. Vapor-transport cooling starts when the metal has cooled down enough so that the vapor film is no longer stable and wetting of the metal surface occurs. This is the fastest stage of cooling. Liquid cooling starts when the surface temperature of the metal reaches the boiling point of the liquid so that vapor is no longer formed. This is the slowest stage of cooling.

HEAT TREATMENT OF NON-FERROUS MATERIAL

Non-ferrous materials, especially the forgings, are heat treated to acquire additional properties or to improve existing properties.

Heat Treatment of Copper and Copper Alloys

Pure copper and some copper alloys can be work hardened and strengthened. The original grain structure of metal completely changes at 590°C; this happens without any appreciable growth in gains. Hence copper and its alloys can be annealed to significant ductility. They can also be annealed by heating to about 590°C (1,100°F) and holding at this temperature for long enough to let the heat fully penetrate the core of the material. It is then allowed to cool to room temperature without any regulation of the cooling rate.

Brass too can be annealed in a similar way; by heating to its recrystallization temperature of 590°C and cooling it to room temperature. The rate of heating or cooling has no apparent effect on the grain size of the metal.

Bronze can be similarly heat treated to correct the hardening effect of cold working. The metal is heated to the recrystallization temperature and cooled to room temperature.

Heating to above the recrystallization point and rapid cooling hardens aluminum bronze. *This hardens and strengthens the material, making the hardened aluminum bronze best suited for tools that are to be used in no-spark areas. These materials are often specified to be used in inflammable and explosive environments.*

Heat Treating Aluminum and its Alloys

Cold worked aluminum alloys acquire strain hardness, requiring the material to be annealed. The recrystallization temperature of alloys may vary from 340 to 400°C (650 to 750°F) the temperature and the rate of heating is specific to the alloy and must be carefully applied.

Sometimes aluminum-copper alloys are also solution heat-treated, although this is dependent on the changing solubility of aluminum for the constituent of $CuAl_2$ during heating and cooling of the material in a solid state. If an alloy containing up to 4.5% copper is reheated to a temperature of about 510°C (950°F) for a considerable time (up to about 14 hours) most of the $CuAl_2$ present in the annealed alloy will dissolve. When this solid solution structure is quenched rapidly in water, most of the $CuAl_2$ will remain in solution in the metal. This treatment improves the corrosion resistance of this alloy and increases its tensile strength.

HEAT-TREATING FURNACES

Various types of furnace are used for heat-treating steel. The type of treatment required, the size and shape of the equipment or parts, the volume of the work, the production output and economic factors all need to be considered in determining what type of furnace is to be used for a specific job.

Batch type or continuous type furnaces are often selected where the volume of production output is large. These furnaces are equipped with temperature control devices. They also control the atmosphere of the furnace chamber, which give such furnaces a distinct advantage in producing good quality output. The handling of the charge, extraction from the chamber and moving to the next stages, e.g., cooling cycle, or moving to the quenching bath, are often mechanized.

Furnaces may be oil fired, gas fired or electrically heated. They may be further classified as follows:

Direct-fired furnace: In this type of furnace the steel is in direct contact with the hot gases of combustion.

Semi-muffled furnace: In this type of furnace, the gases are deflected from the parts and only the heat is allowed to circulate around the parts being heat-treated.

Muffled furnace: In this type of furnace, the heating gases are burned in a separate chamber and only radiant heat is allowed to come into contact with parts. This furnace has better control over the heating environment around the parts.

Electric furnaces are muffled types, in which resistors contained in a muffler do the heating. The heat is radiated to the steel in a controlled environment in which it is essential to protect the steel as well as the heating sources from oxidation damage. The temperature is controlled and monitored very efficiently, as the heating source is much more stable than in the gas or oil heating furnaces. Oxidation, scale formation and decarburization in the furnace environment can also be controlled by creating an artificial environment through the introduction of gases. These are commonly a mixture that is predominantly reducing, containing carbon monoxide (CO) or hydrogen (H_2). The key advantage of an electric furnace is that the parts come out with clean surfaces, free from oxide scales, and also the temperature control is much better.

Mechanical attachments which improve the furnace performance are the rotary hearth, rotary drum belt conveyor, and tilting heath mechanism. Because of better control of temperature and environment these furnaces are often used as brazing furnaces.

LIQUID HEATING BATHS

Liquid baths are useful in routine hardening and tempering operations. It is quick, provides uniform application of heat to all parts being heated, and can be used as a batch furnace for a number of components at a time. Its other advantages are clean heating without surface oxidation, and control over any possible distortion of material being heat-treated. Generally there are three basic types of bath.

Oil baths are used to heat steel for tempering treatment up to 315°C (600°F). A higher temperature is difficult to achieve, due to the flammability of the oils, and also at higher temperature the oil residue is very difficult to remove, and hence counters the main objective of producing clean parts.

Lead baths are capable of achieving a higher temperature, ranging from 340°C to 870°C (650°F to 1,600°F), and are suitable for tempering high-speed steels and alloy steels. Lead oxidation is a major issue with these baths, and it must be protected from oxidizing by using molten salts and charcoal. Steel dipped in a lead bath must be protected from what is referred to as 'lead sticking' by the application of a thin layer of salts.

Salts baths are useful for a variety of treatments for steel and other alloys. Steel can be heated in a salt bath for annealing hardening or tempering. This is possible because salt baths that have greater range – from 150°C to 1,300°C (300°F to 2,400°F). Salts react dangerously in several environments, and some, such as cyanide salts, emit dangerous fumes. This requires proper health and safety precautions to be taken when working in these environments.

Because salts do not have universal reactions with different materials, their selection is specific to the metal and alloy. A salt that is neutral to the metal must be selected.

Salt baths are often used for heat treatment of aluminum and its alloys to allow better controlled and more uniform heating. Sodium nitrate baths are common practice for aluminum alloy heat treatment.

Welding Metallurgy and Welding Processes

Introduction

There have been discussions and sometimes arguments about whether welding is an art or a science. Mundane as it might appear, the question is pertinent and in my experience some well-meaning experts often miss the point as to which part of the term 'welding' they are referring to support their arguments. Welding as the physical and practical process of joining two materials in most part is an art, however the study of the heat and melt flow solidification, and the prediction of material behavior under heating and cooling cycles associated with welding is a science. Hence welding is both the art and science of joining metals by use of adhesive and cohesive forces between metals. Welding, brazing, and soldering are joining processes which produce metallurgical bonds.

Both process metallurgy and physical metallurgy, as discussed in previous sections of this book, are involved in welding. Welding is a unique metallurgical activity as it involves a series of metallurgical operations similar to metal production, like steelmaking and casting, but in a rapid succession and on a very small scale. Generally the thrust of the study is on the material's behavior during the application of localized heat, and on the physics of cooling and solidification.

Welding is often compared in a very rudimentary way with casting, because in welding a volume of molten metal is solidified (cast) within the confines of a solid base metal (mold). The base metal may have been preheated to retard the cooling rate of the weld joint, just as in casting the molds are preheated to slow down cooling and reduce 'chilling' of the casting. Upon solidification, the weld deposit or casting can be directly put into service, as welds are often used in as-welded condition – or they may be heat-treated or worked on as required. However such comparison is not an accurate depiction of the welding process. For example, in welding the base metal 'mold' is part of the weld, unlike the mold of a casting, which is removed after solidification, so unlike the casting process, what happens to the 'mold' is of significance in welding. Also, in contrast to casting, in welding the solidification and the nucleation of the weld metal takes effect on the basis of the base metal grain structure that is immediately adjacent to the molten metal of welding, and a unique set of metallurgy is created in the base metal that is heated to above austenitic temperature range. This small band of base metal is called the heat affected zone (HAZ). We shall discuss this phenomenon in more detail in this section of the book.

Welding involves a small area relative to the base material of the entire structure. Thus a weld is a very small mass of metal, and is mostly confined to two metals that are heated very rapidly by intense heat, and then cooled rapidly. The dissipation of heat takes place via all three modes; conduction, radiation and convection. Often the large surrounding mass of colder base metal is heated by conduction, which is the major source of heat transfer from weld. The heating and the cooling process after welding are dynamic; equilibrium conditions are seldom seen in conventional welding operations, and generally welding conditions represent a great departure from equilibrium. This is the reason that weld zones often display unusual structures and properties within the confines of the very small area affected by the welding process.

It is thus important that a welding engineer has very good understanding of 'heat' in welding. An understanding of heat generation and the physics of welding are important steps in making a good welding engineer. In the next chapter we start with a basic understanding of the physics of welding. In that we will review the process of heat generation, which is an essential part of the welding processes, along with the cooling, nucleation and solidification physics of molten weld metal and its effect on the parent metal. We will also review heating rate, peak temperature, HAZ and the changes these bring in the properties of weld metal and the metal being welded.

Welding is carried out on the basis of a well thought out and specific plan in order to attain the required material properties. Many regulatory and industrial specifications have well developed plans in this respect. Such plans are called welding procedures. The following is a brief discussion of welding procedures and their role in welding applications.

WELDING PROCEDURES

A welding procedure is a statement of execution; a specific plan prepared by the welding contractor. The procedure details the various variables associated with the proposed welding process, giving an assurance that the resulting weld will guarantee that the required mechanical and metallurgical properties will be met. Any format of form may be used to develop a welding procedure, as long as it gives the essential details. Some international specifications that address welding requirements have developed a format for this purpose: AWS D1.1 has the E-1 form for pre-qualified procedures, similarly ASME Section IX of Boiler and Pressure Vessels Code has a set of sample forms, QW-482 for Welding Procedure Specification, QW-483 for Procedure Qualification Records (PQRs), and QW-484 (A&B) for Welder Performance Qualification (WPQ) and Welding Operator Performance Qualification (WOPQ).

The plan details all essential and non-essential variables that are important to achieve the required quality of weld. These variables are welding process specific. In ASME section IX, these variables are listed for each welding process, subdivided into essential, supplementary essential and non-essential variables. However these variables are not specific to ASME but are in general agreement with welding technology.

Essential variables are those to which a change, as described in the specific variables, is considered to affect the mechanical properties of the weldments; hence any change in any essential variable will require requalification of the welding procedure. The supplementary essential variables are required for metals for which other sections specify notch-toughness tests and are in addition to the essential variables for each welding process.

The non-essential variables, on the other hand, are those to which a change, as described in the specific variables, may be made in the WPS without requalification.

Some special processes, such as corrosion-resistant and hard-surfacing weld metal overlays, may have different additional essential variables. Only the variables specified for special processes will apply. A change in the corrosion-resistant or hard-surfacing welding process requires requalification.

The correct electrode diameter is one of the variables, which, when used with the proper amperage and travel speed, produces a weld of the required size in the least amount of time. Selection depends on the thickness of the material being welded, the position of welding in relation to the gravity of the earth, and the type of joint to be welded. The welder's experience is also important, since more skill is required to control the weld puddle in out-of-position welds. Inexperience may lead to poor quality welds that may have defects such as inclusions, lack of fusion, incomplete penetration, or porosities.

Welding current can be either direct or alternating, depending on the process, type of electrode and available power supply and material being welded. DC provides a steadier arc and smoother transfer as well as good wetting action and better out-of-position control. Reverse and straight current polarities are used for specific applications. Reverse polarity produces deeper penetration and straight polarity produces higher electrode melting rates.

Physics of Welding

The physics of welding deals with the complex phenomena that are typically associated with welding. The study involves knowledge of electricity, magnetism, heat, light and sound. While studying them, it is essential that the welding engineer familiarizes himself with, and learns to use the correct terminology; this helps with communicating effectively with fellow welding personnel. The American Welding Society (AWS) has developed a library of these terminologies over the years, AWS A-3.0-94 'Standard Weld Terms and Definitions' which is actively used and understood worldwide. It is recommended that these terminologies are used and referenced while conversing about welding.

Applied Welding Engineering: Processes, Codes and Standards.

For the purpose of this chapter we shall define and explain a few important terms and their meanings as applicable to welding engineers.

HEAT

It is the commonly used term for energy. This energy resides in the kinetic motion of atoms in any substance.

The heat flow challenges that are faced by welding engineers are very complex, and often it is very difficult to develop a meaningful mathematical model for them, unless a complex laboratory environment is easily available. It is essential that the welding engineer understands the physics of heat flow and heat transfer associated with the welding process. Thermal energy can be transferred from one body to another by one of the following three methods:

1. Conduction: this transfer is via contact between two metals. A temperature gradient is necessary for the transfer of the heat by this method.
2. Radiation: unlike conduction, no physical contact is necessary for the transfer of heat by radiation. The actual transfer is through photon emission and capture by the target material.
3. Convection: this method is generally applicable to heat transfer in fluids. In this method of transfer, the fluid motion of hotter masses transfers heat to the cooler parts of the fluid in a cyclic motion until all the fluid has reached equilibrium.

In welding, heat is generally applied at a localized point, or in other words concentrated heats applied at the point of the desired bond. The sources of this heat are various. The high energy density heat source is generally applied to prepared edges of the material to be welded. Thermit, fuel gases, furnace heat, electron beam, laser beams, electrical resistance, electric arc and friction heat are some of the commonly used heat-producing sources for welding. The processes that use an external source of heat are usually identified by the type of heat source.

In welding, the transferred power is the rate at which energy is delivered per unit of time from the heat source to the work-piece, which is expressed in Watts. Thus the energy density is the transferred power per unit area of effective contact between the heat source and work-piece, and is expressed in Watts/meter^2. Heat sources are often qualitatively compared in terms of temperature, which is applicable for Oxy-fuel welding or any of the arc welding process, but such a comparison is not accurate for electron beam or laser-beam welding sources. The evolution of welding processes has, in large measure, been predicated on the development of high energy density heat sources. Energy density is an unambiguous measure of 'hotness', applicable to all sources of heat generated for the welding process.

The transfer of heat energy from welding source to work-piece is a complex process. The energy density of a welding heat source cannot be expressed as a precise number. This is because of the unique conditions of each welding process, as identified below.

- It is difficult to define the precise area of contact between the heat source and the work-piece.
- The intensity of the heat is distributed non-uniformly over the contact area, typically with the intensity concentrated at the center of the weld.

One way of regarding the welding heat energy and its heat flow is to consider two distinct processes of heat transfer.

1. Heat is transferred from the source of heat to the work-piece by convection and radiation.
2. On the work-piece the heat is transferred by conduction to the adjoining area.

The heat generated by various welding processes and the arc efficiency of some of the common arc welding processes are tabulated in this section.

DETAILS OF THE HEAT-FLOW IN WELDING

Fundamental to the study of welding is the study of heat-flow. In welding, the application of a heat source is called **energy input**. In case of an arc welding, it is arc energy input, defined as the quantity of energy introduced per unit length of weld from a traveling heat source. The energy input (heat input) is expressed in joules per meter or millimeter. This important measure is calculated as the ratio of total input power in Watts to its velocity:

$$H = P/V \qquad (1)$$

Where:

H = Energy input, joules per mm
P = Total power input of heat source (Watts)
V = Travel velocity of heat source (mm/sec).

If the heat source is an arc, to a first approximation:

$$H = EI/V \qquad (2)$$

Where:

E = volts
I = amperes
V = Travel velocity of heat source (mm/sec).

The above relationship is adequate for general recording of the heat input in welding. However if the objective is to make a precise determination of the heat effects of arc on the material being welded, for example for mathematical modeling, then net energy input, H_{net} should be calculated. This is a function

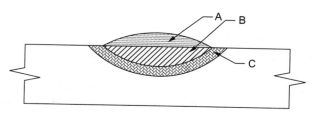

FIGURE 2-2-1 Weld cross-section area and the dilution of two materials in the weld nugget The dilution is calculated as; Dilution = B/ A + B. Note that HAZ C is not part of the dilution calculation. *Picture courtesy of Air force training manual "Basic Welding Technology".*

of heat, source of power input and efficiency of the power source. If the efficiency is taken into the equations (1) and (2) above, then the relationship can be described as:

$$H_{net} = f_1 H = f_1 P/V = f_1 EI/V \qquad (3)$$

Where f_1 is the heat transfer efficiency. This is a ratio of the actual heat transferred to the total heat generated by the heat source. With most consumable electrode arcs, the distinction between H and H_{net} is not important, because the heat transfer efficiency, f_1, is generally greater than 0.8 and often reaches 1.0.

The primary function of most welding heat sources is to melt the metal to be welded. The quantity of metal to be melted is dictated by:

- The size and configuration of the joint
- The number of passes
- The welding process.

The melting efficiency is the fraction of the net energy input, H_{net}, that is actually used for melting the metal. The bead-on plate weld cross-section is shown schematically in Figure 2-2-1.

Note that sections A, B, and C in the sketch are as follows:

A = Area of weld metal, also called reinforcement. This area represents the volume of filler metal added in the molten condition.

B = Area of base metal that is melted.

C = Cross sectional area of the heat affected zone.

The cross sectional area of the weld metal, A_w, is the net area of weld, computed as follows:

$$A_w = B + A \qquad (4)$$

In an autogenous weld (without any addition of filler metal) the cross sectional area of the weld A_w is equal to the area of base metal and is computed as:

$$A_w = B$$

To melt a given area of base metal, a specified quantity of heat is required. This specific theoretical quantity of heat is Q. The amount of heat Q is required to melt a given volume of metal from cold start. This quantity, Q, is a property of the metal or alloy, and is obtained by adding:

- The heat required to elevate the temperature of solid metal to its melting point
- The heat required to convert the metal from solid to liquid at the melting point; i.e., the heat of fusion.

A reasonable valuation of Q in J/mm^3 can be derived from following relationship:

$$Q = (T_m + 273)^2 / 300,000 \qquad (5)$$

Where T_m is the melting temperature of the metal or alloy in degrees centigrade (°C).

Thus for steel, which has a melting temperature of 1576°C, the quantity of heat required to melt one mm^3 steel is $(1576 + 273)^2 / 300,000 = 11.39$ J/mm^3.

Another important factor is the efficiency of melting. The melting efficiency is a theoretical measure of the minimum arc heat f_2 of a weld pass, which can be determined by measuring the weld metal cross section and the net energy input. The melting efficiency f_2 is an inverse function of the minimum arc-heat required to melt the metal. The value of f_2 is obtained by dividing the minimum arc heat required to melt the metal, by the net energy input:

$$\begin{aligned} f_2 &= QA_w/H_{net} \\ &= QA_wV/f_1P \\ &= QA_wV/f_1EI \end{aligned} \qquad (6)$$

The melting efficiency depends on several factors, the principal among them being:

The welding process: Different welding processes have different heat-producing efficiencies; for example varying arc welding sources have varying arc efficiencies, and as a result they have different heating abilities. The arc efficiency of some of the most common welding processes is given in Table 2-3-2.

High-energy processes are rated at up to 100% efficient. Electron beams and laser beams are examples of high-efficiency processes. In these processes, the delivered heat energy is so localized that melting takes place before any heat is conducted away. For welding, a maximum energy density of 10 kW/mm^2 or 76.5 MW/ in^2 can be used; higher intensities would cause boiling and burning. The higher energy output of the electron beam and laser beam processes is utilized for cutting and drilling metal.

The submerged arc welding (SAW) process is a relatively efficient process. Its energy density and the melting efficiency is the highest of all open arc-welding processes.

Simplifying the equation (6) above, by substituting for H_{net} from equation (3), a new relationship between weld metal cross section, A_w and the energy input is arrived at:

$$A_w = f_2 H_{net} / Q = f_1 f_2 H / Q \qquad (7)$$

For a given welding process, there is not much variance between the heat transfer efficiency (f_1) and the melting efficiency (f_2) with changes in specific welding variables like arc voltage, current or travel speed. This highlights another constant relationship; that is the cross section (A_w) of a single pass weld is roughly proportional to the energy input.

The material being welded: A material's thermal conductivity is inversely proportional ($y = k/x$) to its melting efficiency. The effect of thermal conductivity is more pronounced with a low energy density heat source. A good example of this is the low heat energy transfer to work piece. Only 2% of the energy is transferred when welding aluminum with the Oxy-fuel welding process. Thus, metals with higher conductivity are seldom successfully welded with low energy processes.

Joint configuration and plate thickness are the other factors which affect heat-flow in welding. These factors contribute directly to the determination of pre-heat temperature for welding some materials; for this the term relative thickness is used, indicated by letter τ in calculations. The calculation is made using a relationship of base metal thickness, volumetric specific heat of material, the temperature differential and net heat input:

$$\tau = \{h \sqrt{P^c (Tc - To)}\} / H_{net}$$

where:

τ = The relative thickness of the plate
h = The volumetric specific heat of the material
P^c = Volumetric specific heat ($0.0044 \, J/mm^3 °C$ for steels)
Tc = The temperature at which the cooling rate is being calculated
To = The original temperature of the material

HEAT IN ARC WELDING PROCESSES

In the previous section, we discussed power and efficiency. In that discussion we took an empirical view of the welding process, and now we will explore these factors in more detail, emphasizing the heat process and efficiencies in welding processes.

A large number of welding processes use electric arc as the source of heat for fusion, because of its concentration and ease of control. An electric arc consists of a relatively high current discharge sustained through a thermally ionized gaseous column called plasma.

The power of an arc may be expressed in electrical units as the product of current passing through the arc and voltage drop across the arc. Given a typical current of 250 A, and a voltage of 25 V, the power generated is 6,250 Watts; about 6,250 J/s or 5.93 Btu/s of heat. As stated above, not all the generated heat is utilized in welding; there is significant loss of heat caused by spatter and heat dissipation by convection, conduction and radiation. The effective use of the generated heat is called the efficiency of the process. The efficiency is expressed as a percentage of the generated heat against that actually utilized. This can range from 20% to 85% in various arc-welding processes.

The welding processes GTAW, SMAW and SAW are listed in ascending order of efficiency. In other words the GTAW is the least efficient process and SAW is the most efficient process – utilizing up to 85% of the generated heat.

Arc efficiency is an important factor in the study of weldability, especially when slower travel speed processes are the subject of study. However as the travel speed increases, the following important changes occur:

- The efficiency of heat transfer in the fusion zone increases
- For a constant arc energy input system, the volume of the fused metal increases

As a consequence of the above changes, arc energy input is no longer an adequate parameter to use as a factor for measuring weldability in automatic welding processes associated with process characteristics of higher speed and higher current.

HEAT IN PLASMA ARC CUTTING AND WELDING

In plasma arc, the heat source is the arc which is forced through a nozzle to constrict its diameter. A higher voltage is used to provide this extra force. When the forced arc is passed through the constricted passage of the nozzle, a smaller diameter of the arc column is created. This results in the following:

1. A higher density (intensity) arc is created.
2. The temperature of the arc is increased.

86 kW of power passing through a 3.2 mm nozzle with a gas flow of 70 liters/min will produce a power density of approximately 8.5 kw/mm^2 and an average gas temperature between 9,700°C to 14,700°C (16,232°F to 26,492°F) can be achieved. At these temperatures, the resulting gas velocity would reach sonic levels. These properties are good for plasma arc cutting. However, for a Plasma Arc Welding (PAW) process, the control of the molten pool is important, hence lower gas velocities are used. More details of the process are discussed in the later parts of this book in the sections on welding processes.

HEAT IN RESISTANCE WELDING

The resistance welding process uses a combination of mechanical force and heat to accomplish welding. Electrical resistance between the two work-pieces generates the heat. Heat is also generated from the electrical energy of the welding current flowing through the work-piece. The work-pieces are generally connected to the secondary circuit of a transformer that converts high voltage, and low current of commercial power into high current and low voltage as suitable welding power.

The heat generated by current flow may be expressed by:

$$H = I^2Rt \tag{8}$$

Where:

H = Heat generated, joules
I = Current, amperes
R = Resistance, ohms
t = Time of current flow in seconds

The resistances that are important in these welding processes are composed of several parts:

- Contact resistance between the electrodes and the work
- Contact resistance between the work-pieces
- Body resistance of the work-pieces
- Resistance of the electrodes.

The heat requirement is calculated as follows:

H in joules = (current in Amperes)^2 × (material thickness in meters)
　　　　　　× (Time in seconds, duration of current applied)

Approximately 1,381 Joules (about 600 btu/lb) are required to heat and melt 1 gram of steel.

Let us consider another example, in which we use a welding machine that has a heat output of 1,000 J. Assuming that the fusion zone of a weld is a cylinder of 5 mm diameter and 1.5 mm height, the fused metal would have a volume of approximately 31 mm^3 and a mass of 0.25 gram. The heat required to melt this mass is 345.45 Joules, and the remaining heat (1,000J − 345.45J = 645.55J) will be absorbed by the surrounding metal.

However, if a capacitor-discharge power supply unit is used for making a projection weld between two steel sheets that are each 1 mm thick, the current pulse used is 30,000 A and the weld time is 0.005 seconds. In this case the calculation for the required heat (H) will be as follows:

$$H = (30,000)^2 (0.0001) (0.005) = 450 \text{ J}$$

The smaller amount of heat required in this case is due to the lower heat loss and localized heat application, due to the localized weld interface.

HEAT IN ELECTROSLAG WELDING (ESW)

In the Electroslag welding (ESW) process the electrode is fed through an electrically charged bath of molten slag. The resistance of the slag bath to the flow of current produces heat. The majority of that heat is concentrated in the slag immediately surrounding the electrode tip. The quantity of heat, H, produced in the slag pool, can be expressed as:

$$H = Eit \tag{9}$$

Where:

E = the voltage, volts
i = current, amperes
t = time, in seconds

The manner in which the heat is transferred in this process is complex; however it is an essential tool to control the quality of the weld. The shape of the weld pool is determined by the following variables:

• The depth and width of the slag
• Electrical variables
• Geometry of the weldments, and
• The dimensions of the electrode and base metal.

In processes where slag bath has a very important role in the flow of welding current, the properties of the slag are an important variable for good weld heat. The slag should have:

1. Stable conductivity in heat and range of current
2. Stability at the operating temperature ranges
3. Low volatility at operating temperatures
4. Suitable chemical reactivity
5. Suitable viscosity, for stability.

ESW can be stabilized over constant slag temperatures, by maintaining a balance over the heat generated and heat lost. The geometry of the cross section of the weld pool should be shallow, with a large radius of curvature. A shallow pool promotes vertical freezing, which allows an acute angle to develop between weld metal grains. High current and low voltage promotes shallow pools. Conversely, deep pools are produced by high voltage, high wire feed and low currents. The slag bath should also be shallow to avoid incomplete fusion in the weld.

Both AC and DC currents are used for the ESW process. The choice depends significantly on the slag/metal reaction, in contrast to the operating characteristics associated with other welding processes.

When direct current electrode positive (DCEP) is used, a constant potential power source and wire feed technique are needed.

Alternative current (AC) for similar weld quality will require higher voltage. This will result in higher welding heat as calculated using the H = Eit relation discussed above.

HEAT IN WELDING PROCESSES USING CHEMICAL SOURCES

Chemical energy stored in a variety of forms can be used for welding, by converting chemical energy into heat energy. In general, these can be classified as either Oxyfuel welding or Thermit welding processes.

Oxyfuel welding requires two fuel characteristics:

1. A high flame temperature capable of melting and controlling the molten weld metal.
2. A neutral or reducing atmosphere surrounding the molten pool to prevent contamination of the weld metal before solidification.

Most commonly used fuel gases, when mixed with oxygen, achieve temperatures up to 2,760°C, (5,000°F). At its maximum temperature the flame is oxidizing, hence it is not suitable for welding as they will produce oxides and contaminate the molten weld metal.

Table 2-2-1 lists various oxyfuel gases and the maximum temperatures they produce in reaction with oxygen.

From the table we can see that only the oxy-acetylene (C_2H_2) welding (OAW) flame is capable of developing enough heat to melt steel. MAPP can also be used for welding, with some additional precautions.

Acetylene combustion in oxygen takes place in two phases. In the first phase, carbon is burned producing carbon monoxide, and the remaining hydrogen unconsumed. The burning takes place in the small, bluish-white cone of the flame where the gases mix; this reaction provides the heat that is most effective for welding. In the second phase, the carbon monoxide is converted to carbon dioxide, and hydrogen to water vapor. This takes place in the large blue flame surrounding the welding operation, and the heat of this region contributes to the pre-heat of the surrounding metal. The chemical reactions of the two phases of the flame are as follows:

$$C_2H_2 + O_2 \rightarrow 2CO + H_2 \tag{10}$$

$$2CO + H_2 + 1.5\,O_2 \rightarrow 2CO_2 + H_2O(g) \tag{11}$$

In the first reaction, heat is generated by the breaking up of acetylene and the formation of carbon monoxide. This dissociation of acetylene liberates 227 kJ/mol at 15°C (59°F) and the combustion of carbon to form carbon monoxide librates another 221 kJ/mol. The total heat supplied by the first reaction is therefore 448 kJ/mol (about 500 Btu/ft^3) of acetylene.

The second reaction librates water vapor by burning hydrogen, and produces 242 kJ/mol of heat. The combustion of carbon monoxide provides 285 kJ/mol or an additional 570 kJ/mol for the reaction. The total heat supplied by the second reaction is thus 812 kJ/mol; a sum of 242 kJ/mol and 570 kJ/mol. This is equivalent to about 907 Btu/ft^3 of heat.

The total heat supplied by the two reaction phases is 1,260 kJ/ mol (1,407 Btu/ft^3) of acetylene. The first reaction produces about 35.6% of all the heat

---

TABLE 2-2-1 Welding gas flame temperatures

Flame Types	Maximum Temperature	Neutral Flame Temperature °C
Stabilized Methylactylene-propadiene burning with Oxygen gas	2900 °C	2600 °C
Oxygen with Propylene Flame	2860 °C	2500 °C
Oxygen and Propane Flame	2780 °C	2450 °C
Natural Gas burning with Oxygen	2740 °C	2350 °C
Oxy-Acetylene Flame	3100 °C	3100 °C

Table copied courtesy of IAF Basic Engineering Training Manual, 1969.

---

in a very concentrated form. The remaining heat is developed by the outer, feather-like part of the flame, which acts like a pre-heat temperature, thus reducing the steep thermal gradient caused by welding.

Thermit Welding

Thermit welding (TW) is a process that uses heat from an exothermic reaction to produce coalescence between metals. The name is derived from 'thermite' the generic name given to reactions between metal oxides and reducing agents. The thermite mixture consists of metal oxides with low heats of formation and metallic reducing agents which, when oxidized, have high heats of formation. The excess heats of formation of the reaction products provide the energy source to form the weld.

Let us consider an example; if fine aluminum particles and metal oxides are blended and ignited by an external heat source, the aluminothermic reaction will precede according to the following equation:

Metal oxide + aluminum → aluminum oxide + metal + heat

The most common thermit welds reactions are listed in the following table:

$$\tfrac{3}{4} Fe_3O_4 + 2Al \rightarrow \tfrac{9}{4} Fe + Al_2O_3 \quad (\Delta H = 838kJ) \tag{12}$$

$$3FeO + 2Al \rightarrow 3Fe + Al_2O_3 \quad (\Delta H = 880kJ) \tag{13}$$

$$Fe_2O_3 + 2Al \rightarrow 2Fe + Al_2O_3 \quad (\Delta H = 860kJ) \tag{14}$$

$$3CuO + 2Al \rightarrow 3Cu + Al_2O_3 \quad (\Delta H = 1\,210kJ) \tag{15}$$

$$3Cu_2O + 2Al \rightarrow 6Cu + Al_2O_3 \quad (\Delta H = 1\,060kJ) \tag{16}$$

The theoretical estimated maximum temperature of the reaction listed above is 3,200°C (5,800°F). In practice however, the heat ranges between 2,200°C (4,000°F) and 2,400°C (4,350°F). The ignition temperature of the thermit granules used for welding is about 1,200°C (2,200°F), therefore it is safe from fire hazards if stored away from open heat sources.

HEAT GENERATED BY MECHANICAL PROCESSES

The welding processes in which the required heat is generated by a mechanical process are a unique group, comprised of the following:

1. Friction welding.
2. Ultrasonic welding.
3. Explosion welding.

We will now discuss briefly all three welding processes.

Friction Welding

Friction welding (FRW) is a process that bonds a stationary member to a rotating member by developing frictional heat between the two surfaces. The frictional heat is generated by application of forces on the faying surfaces. As far as possible the rotating member is maintained at a symmetrical axis, while the non-rotating member has no restriction on its geometry.

The weld is done by heat generation through the rotational speed and simultaneous application of normal force. Time is a factor in developing heat, and application of force is required to produce a bond. The amount of heat generated is a function of both these variables. It is significant that the radial temperature distribution is non-uniform. The highest heat is on the outer surface where the speed is highest.

In this process, the average interface temperature is below the melting temperature of either member being joined. The bond is metallurgical, and is achieved by diffusion rather than by fusion. This allows joining of dissimilar metals, particularly those metals that, if melted by conventional welding, would form undesirable phases reducing their weldability or making them practically 'un-weldable'. The width of diffusion can vary significantly; in some cases the diffusion line is difficult to define, but in others it is easily detected by low power magnification.

In practice, this process utilizes one of the following methods to generate heat:

1. Relatively slow rotation and high normal force.
2. High speed and relatively low normal force.

3. Using an inertia process in which a rotating flywheel is disengaged from a rotating drive before the start of welding.

The Welding Institute, in Cambridge, UK (www.twi.co.uk) has carried out some pioneering research and development in this field.

Ultrasonic Welding

Ultrasonic welding, (USW), uses high frequency vibratory energy in the weld zone to effect the weld. The use of this process is often limited to spot, straight or circular welds on members of which at least one is of foil thickness.

The process produces a metallurgical bond, in which on the secured members the vibratory energy is transmitted through one or both sonotrode tips which oscillate in a plane essentially parallel to the weld interface. The oscillating shear stress results in elastic hysteresis, localized slip, and plastic deformation at the contact surfaces, which disrupt the surface films and allow metal-to-metal contact. The metallurgical bond produced can join both similar and dissimilar metals without melting. The temperature can reach up to 35 to 50% of the absolute melting point of metals being joined.

The process involves a 60 Hz electrical power that is transformed into high frequency power through a frequency converter. The output ranges between 15,000 to 75,000 Hz, but frequencies outside this range can also be used. Transducers, while maintaining the frequency, achieve the acoustical conversion of the high frequency electrical power. Suitable acoustical couplers are required to transmit the energy to the sontrode tip and into the metals being welded.

The amount of acoustic energy required for welding a given material increases with the material's hardness and thickness. The relationship for an ultrasonic spot welding can be expressed as an approximation that can later be fine-tuned and adjusted to meet the final requirements. The empirical equation used for most common metals such as aluminum, steel, nickel, and copper in thickness up to 0.8 mm is:

$$E_a = K_1(Ht)^{3/2} \qquad (17)$$

Where:

E_a = acoustical energy, J
H = Vickers micro indentation hardness number
t = thickness of the material adjacent to the ultrasonically active tip, in millimeters or inches
K_1 = Constant; 8,000 for 't' in millimeters and 63 for thickness 't' in inches.

The process requires experimentation and trials to establish the precise machine settings needed to produce satisfactory welds in a given material and thicknesses.

Explosion Welding

Explosion welding, (EXW), uses detonation of an explosive to accelerate a component, called a flyer, to high velocity as it collides with the other component being welded. At the moment of impact, the kinetic energy of the flyer plate is released as a compressive stress wave on the surfaces of the two components, forming a metallurgical bond. In this process the collision progresses across the surface of the plates being welded, forming an angle between the two colliding components. The surface films are liquefied, scarfed off the colliding surfaces, and jetted out of the interface leaving perfectly clean, oxide-free surfaces. Under these conditions the interaction within the molecules and atoms and the resultant forces create cohesion, and the resultant weld is without any HAZ.

As described above, the explosion provides the energy for the weld, so the detonation velocity should be within the limits to produce the required impact velocity and angle between the two components. The sonic velocity within the material being welded limits the explosion velocity. This is a very important variable, because if the detonation velocity exceeds the metal sonic velocity, shock waves will be formed. Since these have sonic velocities, they travel faster than the detonation wave, causing spalling along the edges and fissuring at the weld interface. Controlling the amount of charge per unit area can alter the flyer-plate velocity. When high velocity explosives are used, the process will demand thick buffers between explosive and the cladding plate.

The effect of high velocity explosives can be summarized as:

1. Larger waves are produced with same angle of incidence.
2. The range of angles within which waves are produced also increases.
3. The angle of incidence at which weld waviness occurs begins to increase.
4. There is increased tendency for the formation of intermetallic compounds in the weld interface area.

The correct amount of explosives should be used. Too little explosive will be unable to develop a velocity capable of welding.

The explosives used are varied, and include plastic flexible sheet, cord, pressed, cast, granulated and liquid. Standard commercial blasting caps achieve detonation.

The detonation velocity tends to be constant throughout the explosion. The energy released by explosives is dependent on the thickness of the explosive spread and degree of confinement. These factors, along with the ingredients in the explosive used, become the variables controlling the detonation velocity.

HEAT BY FOCUSED SOURCES

Two welding process which use focused energy are laser beam (LBW) and electron beam welding (EBW) processes. The beams are focused to produce heat, and they operate according to the laws of optics.

Laser Beam Welding (LBW)

The laser beams are focused by the arrangements of various lenses. Electrostatic and electromagnetic lenses focus the electron beam. This results in high power densities.

In laser beam welding (LBW), a high degree of spectral purity and low divergence of the laser beam permits focusing on extremely precise areas, resulting in power densities greater than $10\,kW/\,mm^2$ ($6.45\,MW/in^2$). The beam exiting from a laser source may be typically 1 mm to 10 mm in diameter and must remain focused to be useful for LBW applications. The focused spot size, d, of a laser beam is:

$$d = f\theta \tag{18}$$

Where:

f is the focal length of the lens and
θ is the full angle beam divergence

The power density, (PD), at the focal plane of the lens is obtained by:

$$PD = 4P_1/\pi d^2 = 4P_1/\pi(f\theta)^2 \tag{19}$$

Where:

P_1 is the input power
d is the focused spot size of laser beam

Therefore, power density is determined by the laser power, P, and beam divergence, θ. For a laser beam operating in a fundamental mode where the energy distribution across the beam is Gaussian, the beam divergence θ is equal to:

$$\lambda/\alpha \tag{20}$$

Where Greek letter Alpha (α) is a characteristic dimension of laser, and Greek letter lambda (λ) is the wavelength of laser radiation. By combining equations (19) and (20), we find that the PD is inversely proportional to the square of the wavelength of the laser radiation:

$$PD = 4P_1\alpha^2/ f^2\lambda^2 \tag{21}$$

For welding processes, solid-state lasers and gas laser beams are used. Solid-state lasers are single crystals or glass doped with small concentrations of transition elements such as chromium in ruby, or rare earths like neodymium (Nd) in Yttrium-Aluminum-Garnet (YAG) or glass. The industrial gas lasers are carbon dioxide lasers. Ruby and Nd-glass lasers are capable of high-energy pulses, but are limited in maximum pulse rate. Nd-YAG and CO_2 lasers can be operated continuously or pulsed at very high rates.

Most metallic surfaces reflect a significant amount of incidental laser radiation, but in practice sufficient energy is absorbed to initiate and sustain a continuous molten puddle. Since ruby and Nd-glass lasers have high energy outputs per pulse, they can overcome most metal welding reflectivity problems. However, they have a low pulse rate, typically ranging between 1 to 50 pulses per second, which results in a slow welding speed in thin gauge material. Nd-YAG and CO_2 lasers are capable of high speed continuous welding, because they produce a continuous high wave output. They can also be pulsed at several pulses per second.

Electron Beam Welding (EBW)

Electron beam welding (EBW) is another focused beam welding process. The process develops energy by bombarding the work-piece with a focused beam of high velocity electrons. The power density (PD) defines the process's ability to develop enough heat for welding. The PD in Watts per unit area is obtained by the following equation:

$$PD = neE/A = EI/A \qquad (22)$$

Where:

n = total number of electrons per second in the beam
e = the charge on an individual electron (1.6×10^{-19} coulombs)
E = the accelerating voltage on the electrons, V, in volts
I = the beam current, in amperes
A = the area of the focused beam at the work-piece surface

The beam current, accelerating voltage and welding speed are the factors that determine the depth of penetration of a focused beam.

The power concentration of 1 to $100\,kW/mm^2$ is routinely achieved, and up to $10\,MW/mm^2$ can be obtained for most welding. The concentration of energy is dependent on accelerating voltage. Electron beam welding is generally performed at voltages between $20\,kV$ to $150\,kV$; a higher voltage corresponds to a higher power density.

The EBW process has several advantages due to its focused heat source for welding, as listed below:

1. Welds with higher depth to width ratio can be successfully achieved.
2. High strength of weld can be achieved.
3. Ability to weld thick sections in a single pass.
4. The relative low heat input results in low distortions in the base metal.
5. Very narrow HAZ.

The welding process involves the application of high power density, which instantly volatizes the metal. This creates a needle-like, vapor-filled cavity or keyhole in the work-piece, which allows the beam to penetrate through the section of the metal to be welded. The cavity is kept open by the pressure of the vapor. The flow of the molten metal is from the front to rear of the keyhole, as the weld solidifies.

Three commercial variants of the EBW process are given Table 2-2-2.

TABLE 2-2-2 EBW Variants

	EBW Variant Description		Operating Pressure
1	High Vacuum EBW	A pioneer process	13 MPa (10^{-4} torr) or lower
2	Medium Vacuum EBW	Soft vacuum process	At 13 MPa (10^{-4} torr)
3	Nonvacuum EBW		At 100 kPa (1 atm)

OTHER SOURCES OF HEAT IN WELDING

The energy required for diffusion welding (DFW) is unlike arc, resistance, oxyfuel or thermit welding processes, though the heat is supplied by furnace. These furnaces may be electric, gas or chemical.

The diffusion phenomenon is of prime importance in metallurgy because several phase changes occur in metal alloys that involve redistribution of the atoms. These changes occur at rates that depend on the speed of the migrating atoms.

Diffusion welding is a solid state welding process that produces coalescence at the faying surfaces by the application of pressure at elevated temperatures. This process is also known by some non-standard terms, such as pressure bonding, diffusion bonding, self-welding, and cold welding.

Usually diffusion in metal systems occurs by three different processes. These are dependent on the mode of the diffusing element.

1. Volume diffusion.
2. Grain boundary diffusion.
3. Surface diffusion.

Each of these modes has different diffusivity constants. The specific rates for grain boundary and surface diffusion are higher than rates of volume diffusion.

Adolf Fick derived his first law of diffusion in 1855, and this is used to solve for the diffusion coefficient (D). Fick's first law describes the diffusive flux to the concentration field; it assumes that the flux goes from regions of high concentration to regions of low concentration, with a magnitude proportional to the concentration gradient. In one (spatial) dimension, this is given by:

$$J = -D \, (\partial \phi / \partial x)$$

Where:

- J is the diffusion flux in dimensions of [(amount of substance) length^{-2} time^{-1}], for example (mol/m^{2}s). J measures the amount of substance that will flow through a small area during a small time interval
- D is the diffusion coefficient in dimensions of [length^2 time^{-1}], for example (m^2/s)

- ϕ (for ideal mixtures) is the concentration in dimensions of [(amount of substance) length^{-3}], for example (mol/m^3)
- χ is the position [length], for example m (meters).

D is proportional to the squared velocity of the diffusing particles, which depends on various factors like the temperature, the viscosity of the fluid and the size of the particles.

In two or more dimensions we must use ∇, the del or gradient operator, which generalizes the first derivative, by obtaining the value of J as in the following equation:

$$J = -DJ = -D\nabla\phi$$

The driving force for the one-dimensional diffusion is the quantity derived from the following relationship:

$$-(\partial\phi/\partial x)$$

The driving force derived from the above equation is the concentration gradient for an ideal mixture. However, for chemical systems that are not ideal solutions or mixtures, the driving force for the diffusion of each species is the gradient of chemical potential for this species. Then Fick's first law (one-dimensional case) can be written as:

$$J_i = -Dc_i\partial\mu_i/RT\partial x)$$

Where:

The index i denotes the ith species, c is the concentration (mol/m^3), R is the universal gas constant J/(K mol)), T is the absolute temperature (K), and μ is the chemical potential (J/mol).

Using Fick's first law of diffusion, for metals we derive the following basic equation for diffusion:

$$1/A\,dm/dt = -D * \partial c/\partial x \tag{23}$$

Where:

dm/dt = rate of flow (g/s) of metal across a plane perpendicular to the direction of diffusion
D = diffusion coefficient whose values depend on the metallic system being considered (mm^2/s); the minus sign expresses a negative concentration gradient
A = the area (mm^2) of the plane across which diffusion occurs
$\partial c/\partial x$ = the concentration gradient that exists at the plane in question (c is expressed in g/mm^3)
x = distance, mm

The diffusion coefficient (D) is usually not constant. It is a function of such dynamics as temperature, concentration, and crystal structure.

The essential ingredients of the process are time, temperature and pressure. Atomic vibration and mobility increase with increasing temperature. As the faying surfaces are brought into contact by application of pressure, diffusion across the faying surfaces is increased in both directions, resulting in the coalescence required for welding.

Since it is an atomic process, the welding time is longer. The key advantages of the process are low residual stresses, and the ability to fabricate assemblies that require very close tolerance.

Thermal properties of some common weldable materials

Material	Thermal Diffusivity α (m^2/s)	Volume Thermal Capacity Pc (J/m^{3}K)	Thermal Conductivity k (J/m s K)	Melting Point °K	°C	°F
Aluminum	8.5×10^{-5}	2.7×10^6	229.0	933	660	1220
Carbon Steel	9.1×10^{-6}	4.5×10^6	41.00	1800	1527	2780
9% Ni Steel	1.1×10^{-5}	3.2×10^6	35.2	1673	1400	2552
Austenitic Stainless Steel	5.3×10^{-6}	4.7×10^6	24.9	1773	1500	2732
Titanium Alloy	9.0×10^{-6}	3.0×10^6	27.0	1923	1650	3002
Monel 400	8.0×10^{-6}	4.4×10^6	35.2	1573	1300	2372
Inco 600	4.7×10^{-6}	3.9×10^6	18.3	1673	1400	2552
Copper	9.6×10^{-5}	4.0×10^6	384.0	1336	1063	1945

APPLICATION OF THE PRINCIPLES OF WELDING PHYSICS

The following section discusses the application of the principles discussed above in terms of pre-heat, post-heat, rate of heating and rate of cooling, and their effect on weldments.

Pre-Heating

Pre-heating involves welding material that has been heated to a material-specific temperature aimed at reducing the cooling rate, by lowering the temperature gradient. In multiple bead welds, the successive beads may be deposited on metal that has been pre-heated by the preceding beads.

The stresses caused in material by welding can cause serious limits on its application depending on factors like the material's own strength, hardness and structural rigidity. Various stresses that affect weldments are discussed in detail in further chapters of this book. Some of the corrective measures include the control of heat during welding. This is achieved by raising the material temperature to a level that is below its **lower transition** temperature – mostly much below 450°C. This is commonly referred to as pre-heating; for standard carbon steel this temperature would be in the range of 75°C to 150°C. Pre-heat is applied with the objective of reducing the cooling rate of the weld and HAZ of the parent metal. For most carbon steel, a suitably controlled welding process, in combination with pre-heat maintenance, is sufficient to control distortions and post-weld cracking associated with increased hardness, hydrogen diffusion and cooling stress.

Determining the Need for Pre-Heat and the Temperature

Pre-heating, as stated above, is a method for establishing controlled cooling. Metallurgically it is a method of controlling the formations of harmful martensite in the HAZ of steel welds. The formation of martensite, especially in the presence of hydrogen in the material, can be catastrophic. When welding steels that can be hardened, it is important to establish the cooling rate. With the exception of low carbon steel and austenitic steels, most of the steels in construction are hardenable. The temperature at which the cooling rate (Tc below) is calculated is often the temperature at which pearlite is formed; this is usually at the nose of the temperature curve of the material's CCT diagram.

For any calculation of cooling rate or temperature, it is essential to establish whether the heat dissipation is three or two dimensional. This is also termed the relative thickness of the weld, which is dependent on the weld geometry. There are three general characteristics for the heat flow:

1. two dimensional
2. three dimensional and
3. intermediary conditions.

Based on this, the metal may be termed as relatively low thickness, relatively high thickness, or in an intermediary thickness condition. The relative thickness is described by letter τ, and we can use the equations given in the section where we discussed heat flow to determine the relative thickness τ of given weld material:

$$\tau = \{h \sqrt{P^c (Tc - To)}\} / H_{net}$$

If we collect welding data and use the welding current, voltage, and travel speed we can determine the need for and calculate the pre-heat temperature needed to maintain a cooling rate that will reduce the possibility of martensite formation.

For example if we have following welding data;

Voltage (E): 23 V
Current (I): 320 Amps
Plate thickness: 10 mm
Welding process efficiency (f_1): 0.85
To: 25°C (77°F)
Tc: 525°C (977°F)
Travel speed (v): 9 mm/s (20 inch/Minute)
　　We can calculate the energy input H_{net}
$H_{net} = f_1 \, EI/v = 695.11 \text{ J/mm}$

We calculate the relative thickness τ using the given equation;

$$\tau = \{h\sqrt{P^c(Tc - To)}\}/H_{net} = 0.48$$

This is ≤ 0.6, which is indicative of two directional heat flow, hence a thin wall condition exists. To calculate this, the following equation can be used to determine the cooling rate:

$$R = 2\pi k\, pC\, (h/H_{net})^2(Tc - To)^3$$

If the value of τ is somewhere between 0.6 and 0.9 then it may be necessary to evaluate both thin and thick wall equations, and establish the correct values for the desired cooling rate and required pre-heat temperature. It may be noted that the lowest calculated pre-heat value that gives the desired cooling rate must be used.

The following is a case study in which a lifting lug weld cracked after welding.

Weld Toe Crack: Case Study 1

Introduction to the Case

Four can sections of an offshore platform leg sub-assembly were each welded with a lifting lug, also called a pad-eye. On cooling, three of the four welds developed cracks at the toe of the weld.

On an 8+8 jacket with base bottom of about 60 m×30 m, a can section was fabricated in the yard. The subassembly of four cans, each 16 meters high, was fitted with lifting lugs. These lugs were cut from a 30 mm thick ASTM A-36 plate. The cans were 1,067 mm in diameter, pipe that had a wall thickness of 55 mm. The material conformed to API 5L 2H Grade 50, with a nominal composition as given in the table below.

	C	Mn	P	S	Si	Cb	Ti	Al	N	V
Heat	0.18-	1.15- 1.60	0.030	0.010	0.05- 0.4	0.01- 0.04	0.020	0.02- 0.06	0.012	Intentionally not added
Product	0.22			0.015	0.05- 0.45					

Carbon Equivalent Maxima

The carbon equivalent of the material was 0.44, calculated using:

$$CE = C + Mn/6 + (Cr + Mo + V)/5 + (Ni + Cu)/15$$

Mechanical Properties (transverse tests)

Yield strength., ksi (MPa)	Tensile Strength min., ksi (MPa)	% Elongation in 2 in. (50 mm)	% Elongation in 8 in. (200 mm)
51 (352)	90 (620)	26	21

Charpy Impact Properties

The as-rolled slab Charpy V-notch tests performed on three transverse specimens, gave the values reported in the table below.

Due to the low carbon and sulfur contents, the energy of the full-size specimens will often exceed the limits of ASTM E23; therefore, sub-sized specimens were used.

Specimen size mm	Min. Avg. Energy ft-lbs (J)	Min. Single Value ft-lbs (J)	Test Temp. F (°C)
10 × 10	30 (41)	25 (34)	−40 (−40)
7.5 × 10	30 (41)	25 (34)	−40 (−40)
7.5 × 10	23 (31)	19 (26)	−50 (−46)
5.0 × 10	15 (20)	13 (18)	−80 (−62)

Plates were additionally tested to supplementary requirements of S1, S2, S3, S4.

All four pad-eyes were cut from same ASTM A-36 plate that was lamination checked. After cutting and fabrication, and prior to attaching to the can sections, a UT scan was carried out to ensure there were no newly opened laminations.

The abutting sides of the lifting lugs were beveled for a full penetration weld. All four lugs were fitted using tacks welds. These attachments were inspected for satisfactory fitting, and released for welding in accordance with the approved weld procedure already in use in several other constructions in the yard. The qualified SMAW procedure specification (WPS) called for a pre-heat temperature of 100°C, and use of low hydrogen electrodes. The WPS was qualified on a 2.5 inch thick plate, which allowed the WPS to be used to weld up to 8 inch thick plates. Qualified and experienced welders were assigned to the task of welding.

During the work-day, three welds were completed, and left to cool. The next morning, the NDE inspector carried out the mandatory magnetic particle inspection on all three completed welds. All three pad-eye welds had developed toe cracks on the can side of the toe.

A brief investigation was carried out by reviewing the welding inspector's log, to check for compliance with the welding parameters, which was found to be in order. A discussion with the welders revealed that the welds were completed at about 5:45 p.m. the previous afternoon, and after welding they left the weld station and climbed down from the welding perch.

No clear explanation of the crack was established, which was particularly puzzling, since WPS had been successfully used on other jobs in the yard.

The focus was shifted from finding some lapse on the part of the welding crew to a more technical explanation of the cracking. The welding inspector's log was taken as the basis for calculating the cooling rate under the given conditions, given in the table below as "At qualified WPS Pre-heat". It may be noted that the relative wall thickness (τ) is above 0.9 for both 55 mm and 30 mm wall sections. It was established that a thick wall condition exists. The calculated cooling rate at the two welding speeds and hypothetical pre-heat temperatures of 300°C and 250°C ranged from 8.85°C/s to 19.12°C/s.

It was obvious that at this level, the cooling rate of the weld was very fast, and this had caused increased hardness at the thickest section of the weldments. The increased residual stresses (combined with the increased hardness) was exceeded by the cooling stress exerted, causing the weld to crack in HAZ. The hardest points at the toe of weld nearest the thicker section were the fastest to cool.

In an attempt to find a suitable cooling rate of below 6°C/second, a calculation was carried out using a pre-heat temperature of 300°C on a 55 mm can section, and 250°C on a 30 mm thick pad-eye section. It was noted that thick wall conditions still exist, as the calculated τ values are 1.64 and 1.2 respectively, and the resulting minimum cooling rates R are 8.85°C/s and 12.75°C/s respectively. These cooling rates are faster than the desired 6°C/s or slower. To further reduce the cooling rate the pre-heat temperature would need to be increased. Apart from metallurgical considerations, this was neither practical nor possible as it would restrict good performance from the welders, as well as producing logistical issues associated with such precise control of temperature.

It was established that it was not possible to control the formation of hard martensite under the current welding conditions. However it was possible to control the cooling rate by additional measures like either local stress relief or by wrapping the weldments with a heat insulator and to allow slow cooling.

A set of two new procedures were established to supplement the existing WPS, with the pre-heat temperature increased to 200°C. The pre-heat method was also changed from rose-bud propane torches to wrap-around electric heating pads. Two post-welding cooling control options were considered.

1. Immediately on completion of the weld, the weld and the area surrounding the weld were heated to 450°C. On reaching 450°C the temperature was held for an hour, and then gradually reduced at a rate not exceeding 150°C per hour. As the temperature fell to 150°C the weld was allowed to cool overnight wrapped in insulating blankets, in a very controlled manner.

2. After welding, the weld was put on stress relief PWHT, in which the weld area and adjoining material were heated to 650°C, held at that temperature for 4 hours and then allowed to cool slowly.

All regular tests associated with WPS procedures were carried out, including the hardness test on the weld and HAZ, and then on a cross section of the weld and HAZ. Results from both test specimens are given in the following table.

Weld Type	Hardness on Weld Surface Vickers	Macro-Section Hardness $V_{10\,Kg}$		Micro-Structure
	Weld	Weld	HAZ	
Heated to 450°C	242, 239, 210	Cap$^{+1.5\,mm}$: 188, 169, 173	Cap$^{+1.5\,mm}$: 201, 210, 213	Pearlite and islands of tampered martensite
		Mid: 169, 174, 170	Mid: 172, 166, 171	
		Root$^{+1.5\,mm}$: 181, 175, 168	Root$^{+1.5\,mm}$: 177, 172, 181	
Stress relief at 650°C	239, 232, 236	Cap$^{+1.5\,mm}$: 185, 176, 177	Cap$^{+1.5\,mm}$: 200, 198, 209	Pearlite and islands of tampered martensite
		Mid: 176, 164, 178	Mid: 170, 172, 168	
		Root$^{+1.5\,mm}$: 185, 179, 181	Root$^{+1.5\,mm}$: 189, 183, 191	

The micro examination of the weld cross section was also reviewed; some islands of tempered martensite structure were noted in both specimens.

Both versions of the post-weld cooling control method resulted in acceptable hardness and micro-structure in the weld, and were able to reduce the possibility of post-weld cracks. The deciding factor between them was fabrication convenience and logistics of the yard.

Volt	I (Amps)	S (mm/s)			Process Efficiency	Heat Input (J/mm)	Hnet
24	230	4			0.9	1380.00	1242.00
24	230	6			0.9	920	828
Pc	To	Tc	TC-To	h	Tau (τ)		
0.0044	300	550	250	55	1.64	At Maximum Pre-heat	
0.0044	250	550	300	30	1.20	At Maximum Pre-heat	
0.0044	100	550	450	55	2.20	At Qualified WPS Pre-heat	
0.0044	100	550	450	30	1.47	At Qualified WPS Pre-heat	

Cooling rate calculation for relative thick (τ) wall condition as indicated

Π	k	Cooling rate (R°C) at Tc - To	Cooling rate (R°C) at Tc - To
3.1416	0.028	8.85	13.28
		12.75	19.12

Cooling rate calculation if relative thin wall condition was $\tau \le 0.6$

Π	k	Cooling rate (R°C) at Tc - To	Cooling rate (R°C) at Tc - To
3.1416	0.028	23.72	53.37
		12.19	27.44

In the above calculations using two different values for *To* and two different values of Hnet does not give the desired cooling rate of $\le$ 6°C/Second. This will require that a post-weld heat treatment is considered.

Post-Weld Heat Treatment (PWHT)

For some material, pre-heat alone is not sufficient to restore the properties. In such cases, after a weld is made, a post-weld heat treatment may be required to alter the unusual structure and properties produced by the rapid cooling that is associated with welding process. Post-weld heat treatment is carried out to soften the weld zone (annealing) or to reduce hardening. Tempering operations can be performed to obtain weld zone properties that are similar to those of the base metal.

A metallographic microscope allows visual examination of the microstructural changes produced by the thermal effects of the welding operation.

The metallurgical factors important for controlling the important features of a weld joint are:

1. The composition and soundness of the weld metal.
2. The microstructure of the weld metal and heat affected zone of the base metal.

Most welding processes apply heat to the work-pieces and some of the metal in the joint area is melted. In terms of metal properties, the harmful effects of applying heat for welding outnumber the benefits. By controlling the heating and cooling cycles, it is possible to minimize the undesirable effects.

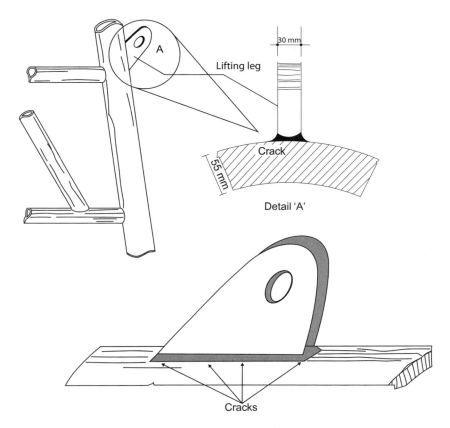

FIGURE C1-1   Location of cracks in the platform leg and lifting lug.

The welding process generates unusual combinations of time and temperature gradient; the temperature changes during welding are wider and more abrupt than in any other metallurgical process. The options of working and heat treatment to restore optimum properties are usually restricted in the case of welded structures, also called weldments.

The following are important points to consider when evaluating the thermal effects of welding:

- Rate of heating
- Length of time at temperature
- Maximum temperature
- Rate of cooling
- Cooling end-point.

As described above, factors like high stress or higher hardness, associated with various grades of steel may require PWHT to make the weldments suitable for the designed service conditions. Such steels require an additional post-weld heat treatment, often called PWHT. The PWHT for steels are often limited to heating to steel's **lower transformation** temperature, i.e., below the austenite start range – that is below 723°C (1,333°F) and cooling slowly. This stress relieving aims to reduce stresses by tempering the few martensites that might have developed during fabrication welding.

Low alloy steels have a **lower transformation** temperature slightly above 723°C (1,333°F) and specific care must be taken to determine the specific **lower transformation** temperature for the specific steel.

Other steels of higher strength and levels of alloying elements may require additional or different heat treatments. Some of these are annealing, normalizing, quenching and tempering. Various heat treatments are discussed in more detail in further chapters of this book.

HEAT AND TIME IN WELDING

Heat will move from one area to another whenever there is a difference in temperature. Heat transfer occurs in three ways, as discussed in the beginning of this chapter, where we discussed heat flow, and the transfer of heat by conduction, convection and radiation.

The difference in temperature per unit distance is called the temperature gradient. Welding involves very steep temperature gradients between the heat source and the work, and within the work-piece itself. The primary mode of heat transfer in welding is conduction, however all three modes are involved to some extent.

Heat Input

Relative to the overall volume and area of the work, the area and volume of the heat-affected zone in weldments is very small. To a significant degree this is effectively controlled by the heat input. The magnitude of the rate of input energy (which is the product of the arc-efficiency and the energy produced per unit by the given welding process, expressed in Watts), the distribution of heat input and the travel speed of the weld are the three variables that determine the heat input in any weld.

Energy Distribution

The density of energy distribution of the heat on the surface of the work is an important factor in heat distribution. The weld and heat affected zone (HAZ) are formed within this heat distribution range, and influence the size and shape of the weld and HAZ. The weld thermal cycle in close proximity to the weld and HAZ are of no significance for some materials, like plain low carbon steel. In such situations, the heat input can be treated as a concentrated heat source, and heat input and weld travel speed are sufficient to determine the thermal cycle.

For arc welding processes, the deposition of heat is a distribution of heat flux on the surface of the weldment. An assumption is made that the heat of the arc is applied in an instant, and then the distribution of heat flux can be mathematically modeled at any point from the center of the arc.

Rate of Heating

The rate of heating of a work-piece depends on how hot the heat source is and how efficiently that heat is transferred to the work. A higher temperature at the source means a steeper temperature gradient between it and the work, and so the heating rate will be faster. To illustrate this point, let us consider three welding processes using different sources to generate heat.

In oxy-fuel welding, the heat is generated in the flame, and the gas molecules transfer their thermal energy to the metal. In arc welding, much higher heating rates are encountered. The arc temperature is much higher than that of an oxyacetylene flame. In addition, the arc is kept in intimate contact with the base metal, so there is efficient transfer of heat. In electric resistance welding, the metal itself is the heat source. Since most of the heat is generated at the contacting work surfaces, there is no effective transfer of heat and a very high heating rate is obtained.

Maximum Temperature

More heat is required to melt a given amount of metal than would appear from the mass of metal involved, because once the temperature is raised in one spot, it spreads and raises the temperature of all adjacent regions. So the heat of welding must be sufficient to not only melt the metal required for welding but also to heat the surrounding metal. Slower heat input rates result in greater amounts of heat required above the amount necessary to just melt the metal.

HEAT GENERATION AND TEMPERATURE DISTRIBUTION – PRACTICAL APPLICATION

The physics of heat generation is discussed above. The following is the practical application of those principles in the real work situation. In arc welding, some heating of the material occurs through resistance to the passage of current through the material as it returns via the ground connection, but the arc produces the majority of the heat for welding.

The heat output of an arc is approximately equal to:

(Arc voltage) × (Arc current)
×(Time in seconds that the arc burns, called Arc-Time)

For example, a covered electrode arc operating at 35 volts and 150 amps for a second, liberates 35 × 150 = 5,250 joules every second. This is equivalent to melting 0.02 pounds of steel in a second.

The energy output that enters the metal ranges from 20% to 85% of the input, depending on other welding conditions, such as travel speed. See the table giving the arc efficiency of various welding processes. Since metals are good heat conductors, the atoms in the metal pass heat along to neighboring atoms very readily. The temperature distribution actually occurs over a cross section that is constantly changing as the heat source moves along the weld.

Time at Temperature

The length of time at a maximum temperature depends on maintaining an even balance between heat input and heat loss. In most welding operations, this balance exists only for a very short period of time.

Usually, the temperature in a local area rises to the maximum and almost immediately begins to fall. It is this very rapid thermal cycling which accounts for the unusual aspects of a heat affected zone.

Cooling Rates

In welding, the cooling rates are even more important than the rate of heating. We have discussed this subject in detail previously in this section. Three features of the weld have the most profound effect on the cooling rate:

1. The weld nugget acts as a heat source.
2. The mass of the base metal represents a heat sink.
3. The base metal temperature establishes the initial temperature gradient.

High current, high-speed welds cool more slowly than low current, low-speed welds. Increased heat input results in slower cooling rates, if other factors are held constant.

Base Metal Mass

The mass of metal in the work-piece, previously deposited weld metal, fixtures, chill bars, etc. all act as heat sinks around the weld nugget.

Even with the same plate thickness, the mass of metal around the weld bead can be changed by depositing the bead on the edge of the plate or at an angle between two plates. The heat supplied to the edge bead can only flow in one direction, while the heat in the fillet weld will flow into both plates.

Only the volume of metal within a 3-inch radius of the weld affects the cooling rate through the important temperature ranges. Base metal further away only affects the cooling rate through low temperature levels.

In this context it may be appropriate to discuss a very important subject – arc strike outside the weld groove. Arc strikes are not desirable, since they cause immense damage to the material, due to the high cooling gradient. This is because of the low area of heat relative to the large area of the base metal. This causes the heat to dissipate quickly and form hard martensite which is often the cause of failure. The effect of arc strike can be determined using the principles discussed in the rate of cooling and pre-heat determination. The temperature raised by an arc strike can be determined by the following modified equation:

$$T = To + \{2Q/pC\ (4\eta\ at)^{1.5}\} - Exp\ [\ -\ R^2/4\ at]$$

Where:

T = Local temperature raised by arc strike
To = Initial temperature
$R = 0$
Q = net heat input = qt_{weld} = $\eta\ E\ I\ t_{weld}$
q = net arc power
t_{weld} = duration of arc strike (weld time)
η = Arc efficiency
E = Arc voltage
I = Arc current
V = Welding speed
a = area affected by strike/second (mm^2/s)

Selection of Consumable Electrode or Wire for Welding: Case Study 2

The choice of welding filler wires or electrode is an important step in getting a compatible weld material. Apart from the chemical and mechanical properties that are often compared to match materials, the available options are not good enough. Factors like the availability of welding processes, follow-up machining, heat treatment, specific working conditions or methods, availability of inspection etc. may present specific challenges. An educated compromise is often a necessity to derive the best possible results. The following real-life case is an illustration of this point.

Background to the Case

In a prominent pipeline project, consisting of 24 inch diameter and 0.980 inch and 1.125 inch wall thickness API 5L X 80 (PSL2) grade pipeline, the challenge was to balance the need for a low defused hydrogen weld and ensure good weld penetration. The pipeline operator had resevations regarding the use of mechanized welding and insisted that only manual or semi-automated processes be used for welding pipes on this project.

Given this limitation, developing a suitable welding procedure was a challenge.

The contractors were comfortable with the use of an automatic welding process, because it would ensure a low hydrogen process and full penetration of the weld. However in the given situation, the choice was made to use a SMAW process with cellulose electrodes to ensure good penetration while keeping the diffused hydorogen in weld within the acceptable limits of less than 4 ml/100 gram of weld metal.

As a pre-bid condition, the bidding contractors were asked to develop and qualify a procedure that would meet the above conditions. Contractors submitted their proposed welding procedures, and two procedures that were considered viable are given below.

1. SMAW cellulose electrode for root pass to ensure good penetration and then follow up with either a FCAW or a GMAW process.
2. The second contractor proposed two versions of the procedure. In both of them, root and hot passes were welded with a SMAW cellulose electrode and followed one coupon with the FCAW and the second with the GMAW process.

Along with the above WPS coupons the contractors were also asked to weld a root and hot pass on a coupon with a cellulose electrode. This weld was used as reference to establish the level of dissolved hydrogen in the weld.

A pre-heat temperature of 120°C was specified and maintained for all welds. Contractors completed all welds. Weld was not stopped until completed except for the normal cleaning and grinding that is part of the welding proceess. This was to allow as much time and temperature as possible for the hydrogen from the cellulose dissolved in the weld metal to escape through the successive low hydrogen weld metal passes.

Among several other standard tests involved with the welding procedure, one more test was added to evaluate the amount of dissolved hydrogen in the weld, especially in the sampling from root and hot pass regions.

The Testing
Samples from all the welds were sent to the laboratory to analyze for diffusible hydrogen in the weld using methods described in AWS A4.3.

Cellulose Weld Metal Testing
The laboratory was instructed to first test the root and hot pass weld metal from the cellulose electrodes. Four specimens were cut and analyzed for the diffused hydrogen level, according to AWS A4.3 (Rev. 2006). A standard mercury displacement method was used for analysis.

The temperature of the bath was maintained at 45°C ± 1°C. The diffused hydrogen was allowed to evolve for 72 hours. The laboratory reported that all the cellulose electrode root and hot pass contained hydrogen in excess of 16 ml of diffusible hydrogen per 100 grams of weld metal.

The average of the three specimens tested was 40.89 ml diffusible hydrogen per 100 grams of weld metal. The report makes it clear that the welds in question have diffusible hydrogen in excess of the acceptable limit of 4 ml/100 g as specified for the rest of the welds in the main pipeline. The reported average diffusible hydrogen in the weld metal is far more than 4 ml/100 g maximum that the engineering team could accept as a low hydrogen weld. This was used as the reference to evaluate the rest of the welds. (See the test report.)

1. The specimen from weld coupons completed with cellulose root and hot-pass plus low hydrogen process fill and cap welds was also similarly tested according to the AWS A 4.3 mercury displacement method.
2. The resulting diffusible hydrogen in each weld was recorded between 3.38 ml/100 g and 4.95 ml/100 g of weld metal. (See the test report.)

TEST REPORT 1

Diffusible Hydrogen Test as per AWS – A4.3 – 93 (R 2006 C)

Test Report No.   :   MHEPL/DH/AWS/08-09/03   Date of Report : 20/05/2008

Sample No.	(1)	(2)	(3)	(4)
Barometric Pressure at the time of analysis (mm of Hg) (B)	761	761	761	761
Ambient temperature at the time of analysis in ^{0}C (Ta)	35	35	35	35
Weight of middle sample + weld metal in grams (M_2)	179.5183	180.4443	180.0822	181.1649
Weight of middle sample before welding in grams (M_1)	172.9622	173.8284	172.9563	173.1536
Weight of weld deposite in grams ($M_2 – M_1$)	6.5591	6.6159	7.1259	8.0113
Hydrogen Volume collected at the end of test in mL (Vg)	4.0	4.0	4.0	4.0
Head of mercury in tubes in mm (H)	144	142	145	148
Hydrogen corrected to STP, mL ($V_{g\ STP}$)	2.8783	2.8876	2.8737	2.8597
Volume of Hydrogen in mL / 100 gms of weld deposite V_{DH}	43.88	43.65	40.33	35.70

Average Diffusible Hydrogen, mL / 100g = 40.89

Calculation : $V_{g\ STP}$ $= \dfrac{V_g(B – H)}{760} \times \dfrac{273}{273 + Ta}$

V_{DH} $= \dfrac{V_{gSTP} \times 100}{M_2 – M_1}$

Diffusible Hydrogen Test as per AWS – A4.3 – 93 (R 2006 C)

Test Report 2

Root and Hot Pass Fill and Cap	Specimen			
	Contractor -A		Contractor -B	
	SMAW Cellulose FCAW	SMAW Cellulose GMAW	SMAW Cellulose FCAW	SMAW Cellulose GMAW
Test Parameters				
Barometric pressure at the time of the test (mm Hg) (B)	761	761	761	761
Ambient temperature at the time of analysis in C (Ta)	32	32	32	32
Weight of the sample with weld in grams (M2)	158.441	159.068	156.989	160.002
Weight of the sample before weld in grams (M1)	153.343	154.876	152.385	154.937
Weight of the weld deposit in grams (M2-M1)	5.098	4.192	4.604	5.065
Volume of hydrogen collected at the end of the test in ml (Vg)	0.26	0.19	0.28	0.21
Head of mercury in tubes in mm (H)	142	142	142	142
B-H	619	619	619	619
273-ta	205	205	205	205
Volume of hydrogen corrected to STP in ml (VgSTP)	0.212	0.155	0.228	0.171
Volume of dissolved hydrogen in ml/100 grams of weld deposited (VDH)	4.15	3.69	4.95	3.38

Diffusible Hydrogen Test as per AWS – A4.3 – 93 (R 2006 C)

Welding and Joining Processes

As we have learned from the introduction to this section, many welding processes require the application of heat and or pressure to produce a suitable bond between the two parts being joined.

The most common conventional and commercially practiced weld process are listed in Table 2-3-1. This table describes the materials that can be welded or joined using the specified process. It may be noted that each processes has

TABLE 2-3-1 Indicates General Limits of Joining/Welding Processes that Apply to the Material Listed in the Left Column

Material	Welding Processes														Other Joining Processes	
	SMAW	SAW	GMAW	FCAW	GTAW	PAW	ESW	EGW	RW	OFW	DFW	FRW	EBW	LBW	B	S
Carbon Steel	x	x	x	x	x		x	x	x	x		x	x	x	x	x
Low Alloy Steel	x	x	x	x	x		x		x	x		x	x	x	x	x
Stainless Steel	x	x	x	x	x		x		x	x	x	x	x	x	x	x
Cast Iron	x	x	x	x						x	x				x	x
Nickel and Alloys	x	x	x		x	x	x		x	x					x	x
Aluminum and Alloys	x		x		x	x	x	x	x	x		x	x	x	x	x
Titanium and Alloys			x		x	x			x		x	x	x	x	x	
Copper and Alloys			x		x	x						x	x		x	x
Magnesium and Alloys			x		x				x			x	x	x	x	
Refractory Alloys			x		x	x			x				x		x	

limitations in relation to the material thickness, and in addition, the cost analysis may not allow for a process to be used on a specific application.

The selection of suitable welding consumable (electrode or wire) is also an important factor in producing suitable weldments. The choice depends on the intended engineering stress on the weld and the welding process being used for the work. Great care is required in this process, and some engineering and commercial compromises are often necessary to arrive at a suitable selection. One example of such compromises and challenges is discussed in the case study included at the end of Chapter 8, Section 3.

The American Welding Society (AWS) standard terms given below are used to describe welding processes. It may be noted that some of these generic process have several variants, and they are used in industries with their new names and acronyms. For instance surface tension transfer (STT) is a GMAW process, and high frequency welding, (HFW), induction welding, seam welding, spot welding, and projection welding are all variants of electric resistance welding (RW). Argon arc welding and heli-arc welding are types of GTAW process. Some of these additional terms are not yet adopted as standard terms by AWS as an independent welding process, but are part of the generic nomenclature.

1. SAW Submerged Arc Welding
2. SMAW Shielded Metal Arc Welding
3. GMAW Gas Metal Arc Welding
4. FCAW Flux Cored Arc Welding
5. GTAW Gas Tungsten Arc Welding
6. PAW Plasma Arc Welding
7. ESW Elecrtoslag Welding
8. EGW Electrogas Welding
9. RW Resistance Welding (EW is also used in some documents)
10. OFW Oxy-Fuel Gas Welding
11. DFW Diffusion Welding
12. FRW Friction Welding
13. EBW Electron Beam Welding
14. LBW Laser Beam Welding
15. B Brazing
16. S Soldering

The term arc welding is used to classify a group of processes that use an electric arc as the source of heat to melt and join metals.

As can be seen from the above discussion and the table, a large number of welding processes use an electric arc as the source of heat for fusion. An electric arc consists of a relatively high current discharge sustained through a thermally ionized gaseous column, called plasma.

The power of an arc may be expressed in electrical units as the product of the current passing through the arc and the voltage drop across it. In a typical example, 350 Amperes at 22 volts will dissipate 7,700 Watts of power. This will generate heat energy of about 7,700 J/s or about 7.30 Btu/s. However, as discussed earlier in this book, it may be noted that not all the heat generated can be effectively utilized for welding, since some heat loss occurs, The effective heat is significantly lower than the generated heat calculated above; this general efficiency of arc is called 'arc efficiency'. The arc efficiency of some very common processes are given in Table 2-3-2. A detailed discussion of power sources for various arc-welding processes is given in the next chapter.

The arc efficiency of a welding process is an important variable that needs to be considered in determining the effect of heat on the weld, and the heat affected zone. The table below lists the arc efficiency range and mean values of some of the most commonly used processes.

Note: the Watt (W) for DC is calculated as;

$$W = EI \tag{1}$$

For AC the relationship is not straightforward, since the phase angle (θ) between the sinusoidal current and the voltage must be considered. This relationship is expressed as $\cos\theta$ and it is called the 'power factor'. Thus for AC, the power (W) will be:

$$W = EI\cos\theta \tag{2}$$

In general, the arc welding process can be described as the process in which a welding arc is struck between the work-piece and the tip of the

TABLE 2-3-2 Process Arc Efficiency

	Process	Arc Efficiency
1	GTAW	Low
2	SMAW	Intermediate
3	SAW	High

Welding Process	Arc Efficiency Factor η	
	Range	Mean
Submerged Arc Welding	0.91–0.99	0.95
Shielded Metal Arc Welding	0.66–0.85	0.80
Gas Metal Arc Welding (CO_2 Steel)	0.75–0.93	0.85
Gas Metal Arc Welding (Ar Steel)	0.66–0.70	0.70
Gas Tungsten Arc Welding (Ar Steel)	0.25–0.75	0.40
Gas Tungsten Arc Welding (He Aluminum)	0.55–0.80	0.60
Gas Tungsten Arc Welding (Ar Aluminum)	0.22–0.46	0.40

electrode. The electrode may or may not be consumed in the joining (welding) process.

Non-consumable electrodes are either carbon or tungsten. The arc welding process may provide the forward movement to electrodes by either manual or mechanical operations.

Similarly, the filler material may be fed into the weld pool by either manual or mechanical means. If all these activities are manual, the process is termed a manual process; conversely if all the processes are mechanical, the process is called automatic. There is the possibility of mixing mechanical and manual operations; in that case the process is called semi-automatic, although some specifications have called them mechanized processes. The majority of semi-automated processes automate the wire-feed. *See the section on semi-automatic welding of pipeline girth weld.*

The mechanization of a process is motivated by the ability to use both a higher current and higher feed speed in some processes, which are interdependent.

Arc welding power supply units reduce the high line-voltage to a suitable output voltage range, usually from 20 to 80 volts. Transformers, solid-state inverters or motor-generators are used as the power source. The same device then supplies the high welding current (30 to 1 500 amps), in either AC or DC, or both.

Constant current machines adjust the load current to maintain a static volt-amp curve that tends to produce a relatively constant load current. The load voltage is responsive to the electrode-to-work distance. These machines are usually used for manual welding with a covered electrode or tungsten electrode, to minimize the inevitable variations in arc length.

Constant voltage machines adjust the load current to maintain a relatively constant load voltage. The load current is responsive to the rate at which a consumable electrode is fed into the arc. This self-regulating machine stabilizes the arc length for wire welding.

In the following section we shall discuss some of the commonly used welding processes and their salient features.

SHIELDED METAL ARC WELDING (SMAW): PROCESS FUNDAMENTALS

The shielded metal arc welding (SMAW) process is one of the earliest arc welding processes, and a versatile one for welding ferrous and several non-ferrous metals. The process uses covered electrodes. An electrode consists of core metallic wire covered with silicate binders, and other materials that may include fluorides, carbonates, oxides, metal alloys and cellulose. The cover is extruded over the wire. The covering is then dried in an oven. The covering has several roles to play:

1. It works as an arc stabilizer.

2. It provides shielding from atmospheric contamination during the molten state by evolving gases, and during solidification by covering the weld metal with slag.
3. It provides scavengers, deoxidizers, and fluxing agents to cleanse the weld and prevent excessive grain growth in the weld metal.
4. It provides a slag blanket to protect the hot weld metal from the air and enhance the mechanical properties, bead shape, and surface cleanliness of the weld metal.
5. It is also a source of alloying elements to produce a compatible weld metal.

How the Process Works

The bare metal end of the electrode is clamped in the electrode-holder, which is connected to the power source. The other lead of the power source is connected to the work terminal. The arc is struck by bringing the electrode into contact with the work surface and then immediately pulling them apart about by 2–3 mm (about 0.08 to 0.12 inches), thus ionizing the gas between the two electrical ends. The resulting arc generates sufficient heat to melt the work metal and the metal electrode. The coalescence of metals is produced by heat from an electric arc that is maintained between the tip of the covered electrode and the surface of the base metal in the joint being welded.

Covered Electrodes Used in the SMAW Process

The American Welding Society (AWS) and several other agencies regulating quality of welding electrodes classify electrodes on the basis of the chemical composition of their undiluted weld metal, or their mechanical properties, or both. Welding current and position are also indicated.

Carbon steel electrodes are included in AWS Specification A5.1. There are two strength levels: 60 and 70 ksi. An example of an electrode designation system is E6010, which is explained below. Some other commonly used electrodes are E7011, E7015, E7018, E7024. For example, in an electrode designated as E6010, the letters and numbers have the following meanings:

- The letter E designates an electrode
- The number 60 signifies that the tensile strength of the deposited weld metal has a minimum of 60,000 psi
- The second to last digit (1) represents the welding position where the electrode is suitable for use (1 = all positions)
- The last digit, (0) refers to the covering type and current type, in this case 0 indicates that the covering is cellulose, and the electrode is suitable for all positions of welding.

Low alloy steel electrodes are included in AWS Specification A5.5. Their numbering system is similar to that used for carbon steel electrodes. A letter or letter/number combination suffix is added to indicate the alloy content (e.g., E7010-A1, E8016-C2). Weld metal strengths of alloy steel electrodes range from 70 to 120 ksi minimum tensile strength. American Welding Society (AWS) specifications use suffixes at the end of the electrode designation to classify the alloying elements, and from time to time reviews new developments and adds or withdraws as required. The following are some of the most common suffixes and their intended meanings.

Suffix	Meaning of the suffix in carbon and alloy steel SMAW electrodes.
A_1	Contains 0.5% Molybdenum (Mo)
B_1	Contains 0.5% Chromium (Cr) and 0.5% Mo
B_2	Contains 1.25% Cr and 0.5% Mo.
B_3	Contains 2.25% Cr and 1% Mo.
B_4	Contains 2% Cr and 0.5% Mo.
B_5	Contains 0.5% Cr and 1% Mo.
C_1	Contains 2.5% Nickel (Ni).
C_2	Contains 3.5% Mo.
C_3	Contains 1% Ni, 0.15% Cr and 0.35% Mo.
D_1	Contains 1.75% Manganese (Mn) and 0.25% Mo.
D_2	Contains 1.75% Mn and 0.45% Mo.
G	0.5% Ni, 0.3% Cr, 0.2% Mo, 0.1% V, 1% Mn (only one of these elements has to meet the requirement to qualify as a 'G' electrode).
L	Controlled elements (for example low carbon)
M	Meets military requirements.
HZ	Meets weld metal diffusible hydrogen requirements.
	H1 up to 15 ml/100 g of weld metal.
	H2 up to 10 ml/100 g of weld metal
	H3 up to 5 ml/100 g of weld metal
	H4 ≤ 5 ml/100 g of weld metal
R	Meets absorbed moisture test requirements.
Numerical	Numerical suffixes following the above listed indicate toughness properties of the weld metal.

Electrode conditioning refers to the storage and handling of covered electrodes to maintain their optimum moisture content. Low hydrogen electrodes, such as E7018, must be maintained in a holding oven set at 150°F to 300°F. Excessive moisture can cause porosity or lead to hydrogen cracking.

Cellulose electrodes (E6010, E6011) are not required to be conditioned, but they do not operate properly if they are dried out and should be stored in a clean, dry place rather than in an oven.

Corrosion-resistant steel electrodes are included in AWS Specification A5.4. Their classification is based on chemical composition, position of welding, and the type of welding current. For example E310-15 is an electrode with nickel and chromium alloying, that is suitable for use in all positions, with DC current.

AWS specifications exist for nickel alloy, aluminum alloy, and copper alloy electrodes as well as cast iron welding, hard-surfacing and overlaying.

Joint Design and Preparation

Although square groove joints are the most economical to prepare, their thickness is limited to about 6 mm (¼ inch). For thicker members, the edges must be prepared to a contour that will permit the arc to be directed to the point where the weld metal must be deposited. Standard 30° bevel, J-groove and U-groove joints are desirable for intermediate thick sections, since they allow access to the root with the least amount of filler metal required. Higher thicknesses may have composite and double bevel preparations for welding by the SMAW process. The selection of the type of joint preparation is a function of several factors, as described in Table 2-3-3.

Fillet welds require little or no preparation and may be combined with groove welds. Minimum stress concentration at the toes is obtained with concave fillets.

When full penetration joints are required and welding must be performed from one side only, backing bars may be used to provide a retaining surface for the first layer of weld metal to be deposited. When there is access to the other side, a backing weld may be made and ground back to sound metal and the rest of the weld may be completed from the second side.

TABLE 2-3-3 Factors Influencing Joint Design

	Key Factors Influencing the Joint Design	Sub-Factors that Cross-Influence the Selection
1	Material	Type, thickness
2	Welding process	
3	Weld design	Access to the weld from both sides or single side
4	Design demands of the completed weldments	Weld strength, weld esthetics, welds

GAS TUNGSTEN ARC WELDING (GTAW): PROCESS DESCRIPTION

The gas tungsten arc welding (GTAW) process uses a non-consumable tungsten electrode which must be shielded with an inert gas. The arc is established between the tip of the electrode and the work to melt the metal being welded and the consumable filler metal is added either manually or by some mechanized process. The inert gas shielding protects the molten metal from atmospheric contamination white it is cooling, and it also provides the required arc characteristics.

The process may use direct current with positive or negative polarity, attached to the tungsten electrode, although in most applications, the electrode is attached to the negative polarity. Alternating current (AC) is also used to produce different effects on the welding.

Argon and helium are the two inert gases used for this process. Choice of gas and type of current and polarity depends on the type of material being welded and quality of weld desired; for example use of helium gas will result in deeper penetration, and if used with DC current then the process provides the deepest available penetration of the weld.

Use of AC current with argon shielding helps remove oxides from materials that have passivation films, like aluminum and stainless steels.

The process uses constant current welding power. High frequency oscillation is generally provided for AC power sources. High frequency attachments with DC process allow for 'no-touch' starting of the arc, which is a distinct advantage for producing high quality welds.

Process Advantages and Limitations

The GTAW process produces superior quality welds that are free from most defects. The welds are free from slag inclusions, and if properly maintained, they are free from any inclusions, including tungsten inclusions. The process can be used with or without filler metal (autogenous), as required. It allows excellent control of the root pass weld penetration. It can use relatively inexpensive power supplies. It allows precise control of the welding variables. It can be used to weld almost all metals. It allows independent control of the heat source and filler metal additions.

Deposition rates for this process are low. Compared to other processes, the welder requires more skill to be able to produce a quality weld. The weld area must be protected from wind and drafts to maintain the inert gas envelope over the weld zone. Tungsten inclusions can occur if the electrode is allowed to contact the weld pool. Contamination can occur if proper shielding is not maintained, or if the filler metal or base metal is contaminated.

Electrodes

The function of the tungsten electrode is to serve as one of the electrical terminals for the arc that supplies the heat for welding. Tungsten has a melting point of 3,420°C (6,170°F). At high temperatures, it is thermionic and emits electrons. It is the cooling effect of the electrons boiling from its tip that prevents the tungsten electrode from melting.

The electrodes are classified according to their purity and presence of alloying elements. The classification is indicated by the color coding on the tips of the electrodes.

AWS Specification A5.12 classifies tungsten electrodes. EWP is pure tungsten. EWTh-2 is alloyed with ThO_2 to improve arc stability. EWZr-1 is alloyed with ZrO_2. Normally, straight polarity is used to provide cooling of the electrode.

The electrode tip configuration influences weld penetration and the weld bead. For DC welding, the tip should be ground to a specific angle with a truncated end. With AC welding, a hemispherical tip is used. Contamination of the tungsten electrode can occur when it touches either the molten pool or the heated filler metal. Improper gas shielding can cause oxidation.

Joint Design

The five basic joints are shown in Figure 2-3-1. Their variations may be used for welding most metals.

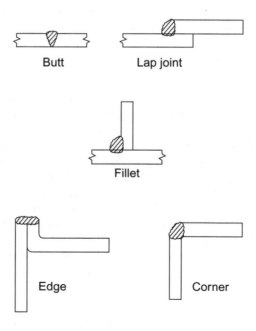

FIGURE 2-3-1   Five basic weld joints.

While designing a weld joint, care must be taken to ensure that there is enough room for proper access by the welder, in order to allow manipulation of the electrode holder so as to obtain adequate fusion of the groove face and addition of filler metal.

The cleanliness of tools used by fabricators and welders for joint preparation is important. Any contamination with abrasive particles or cutting fluids can cause weld defects. Both the filler metal and the base metal must be cleaned to remove all traces of oil, grease, shop dirt, paint, marking crayon, and rust or corrosion.

GAS METAL ARC WELDING (GMAW)

Process Description

Gas metal arc welding (GMAW) uses a continuous (solid wire) consumable electrode. The filler wire is fed by mechanical means, and shielding gas is supplied through the hand-held nozzle, or, if the process is fully automated, then through the nozzles mounted on the mechanized system. After initial setting by the operator, the process of wire feed is self-regulating.

The welding gun guides the feed of the consumable electrode and conducts the electrical current and shielding gas to the work. Energy is provided to establish and maintain the arc and melt the electrode as well as the base metal.

Metal transfer across the arc can occur by any of the following three modes:

1. **Short circuiting transfer** occurs in the lowest range of welding currents and electrode diameters. It produces a small, fast freezing weld suitable for joining thin sections, for welding out of position, and for bridging large

FIGURE 2-3-2a Mechanized welding being performed.

root openings. The wire electrode actually contacts the weld pool at the rate of 20 to 200 times per second. Inductance is used in the power supply to control the amount of heat available before the short circuit occurs.

2. **Globular transfer** takes place when the current is relatively low with all kinds of shielding gas but always occurs with CO_2 and helium. Molten metal drop sizes are larger than the electrode diameter. Because the large drop is easily acted upon by gravity, welding is usually confined to the flat position.

3. **Spray transfer** produces a very stable, spatter-free, axial spray transfer, when the current level is above the minimum transition current. The process can only be used in flat or horizontal positions, since the weld pool is large. Argon or argon-rich gases are required for this process. For welding aluminum, titanium and magnesium and their alloys, an argon-helium mix gas is often used. For welding ferrous material, small amounts of oxygen or carbon dioxide are added to stabilize the arc and eliminate excessive spattering of material. A combination of 75% argon and 25% CO_2 is a gas mix commonly used for carbon steel welding; such gas mixes and other variations are commercially available from gas suppliers. The electrical power used is DC, and the electrode is connected to the positive point of the power source.

Electrode Selection

The filler metal must produce a weld deposit that either closely matches the mechanical properties of the base metal, or provides enhancement to a given base metal property (such as corrosion resistance). The deposit must also be free of discontinuities.

AWS Specification A5.10 includes filler wires for aluminum and aluminum alloy welding. Similarly, A5.19 describes filler materials for magnesium, A5.6 for copper, A5.14 for nickel alloys, A5.16 for titanium alloys, A5.9 for austenitic stainless steels, and A5.18 for carbon steels.

Joint Design

The deep penetration characteristic of spray transfer permits welding in joints that have a smaller included angle, which is advantageous, as it allows for reduced filler metal consumption, and lower labor costs to complete the joint. But such weld joint preparation must be carefully evaluated for potential metallurgical problems which could arise from the high depth to width ratio, which could lead to cracking of the weld.

FLUX CORED ARC WELDING (FCAW)

Process Fundamentals

In the flux core arc welding (FCAW) process, the arc is maintained between a continuously fed filler metal electrode and the weld pool. Shielding can

FIGURE 2-3-2b   Mechanized welding of 48 inch diameter pipe.

be obtained by a flux contained within the tubular electrode; in which case the process is called self-shielding FCAW. The alternative is to supply an external shielding gas; in which case the process is called shielded FCAW (FCAW-S).

The flux-cored electrode is a composite tubular filler metal electrode with a metal sheath and a core of various powdered materials. An extensive slag cover is produced during welding. Self-shielded FCAW protects the molten metal through the decomposition and vaporization of the flux core by the heat of the arc. Gas-shielded FCAW uses a protective gas flow in addition to the flux core action.

The FCAW process combines the productivity of continuous feed wire welding with the metallurgical benefits that are derived from the use of flux and the support of slag. This aids in shaping and protecting the weld bead from contamination, and provides controlled cooling of the weld metal.

Principal Applications of FCAW

The self-shielded method can be used for nearly all applications that would normally be done with the SMAW process, and the gas-shielded (FCAW-S) process covers most of the applications that would use the GMAW process.

The process has higher productivity than the SMAW process; often the chief benefit of FCAW for most applications. Equipment costs are higher, setup and the operation is more complex, and there is a limit on the operating distance from the electrode wire feeder. Applications of this process may be limited by the availability of suitable filler wire and flux combinations for various metals and alloys.

The FCAW process generates large quantities of fumes, which must be removed for the safety of people in the vicinity of the work. Similarly, the slag must be removed between passes to keep the weld free from slag inclusions.

Shielding Gases

Carbon dioxide is the most widely used shielding gas for FCAW, since it is cheap and helps in deep penetration. The CO_2 atmosphere can act as either a carburizing or decarburizing medium, depending on the material being welded. If decarburization occurs (in metals with more than 0.10% carbon), carbon monoxide can become trapped in the weld metal and thus cause porosity.

Gas mixtures are also used to take advantage of their different character-istics. Like the GMAW process, the most commonly used gas mixture that is used for welding carbon steel is the 75% argon and 25% CO_2 commercial mix.

Electrodes

Since a wide variety of ingredients can be enclosed in the tubular electrode, FCAW has good versatility for alloy steels. Alloying ingredients are often included in the core of the wire. AWS Specification A5.20 includes mild steel electrodes, e.g., E70T-1 is an electrode with 70 ksi tensile strength, of tubular construction, suitable for flat and horizontal position welding, with a specific chemical composition. AWS Specification A5.29 includes the low alloy steel electrodes. FCAW electrodes are also specified for surfacing, for stainless steel (AWS A5.22), and for nickel alloys (AWS A5.34).

SUBMERGED ARC WELDING (SAW)

Process Description

The submerged arc welding (SAW) process uses arc to heat and melt the metal being welded. Since the arc is buried under the mound of flux, the arc is invisi-ble to the naked eye. This characteristic gives the process the name submerged arc welding.

The arc is struck between the work and the electrode from the nozzle. This arc provides the required heat for welding. The arc and molten metal are bur-ied under a blanket of granular fusible flux which drops continuously through a chute onto the work. The filler metal is obtained from the electrode and sometimes additional alloying elements are also supplemented through the welding fluxes.

The flux's main role is to stabilize the arc, control the mechanical and chemical properties in the weld deposit, and maintain the quality of the weld.

SAW is a versatile, commercial, production welding process capable of making welds with currents up to 2,000 amps using single or multiple wires or strips of filler metal.

The business end of a continuous bare wire electrode is inserted into a mound of flux that covers the area or joint groove to be welded. After the arc is initiated, a wire-feeding mechanism begins to feed the electrode wire towards the joint at a controlled rate. The feeder is either moved manually, or more commonly, automatically along the weld seam. Sometimes the welding head is kept stationary and the work is moved under the welding head; most of the pipe mills use this practice.

The heat evolved by the electric arc progressively melts some of the flux, the end of the wire, and the adjacent edges of the base metal. A pool of molten metal is created beneath the layers of the liquid slag. This pool is in a very agitated state and gas bubbles are swept up to the surface. The flux floats on the molten metal and completely shields the welding zone from the atmosphere.

Materials

Carbon steels up to 0.29% carbon, low alloy steels up to 100 ksi yield strength, chromium-molybdenum alloys up to 9% chromium and 1% molybdenum, stainless steels, and nickel alloys are often welded with the SAW process.

SAW electrodes can produce matching weld deposits by varying the electrode-flux combination. SAW electrodes vary in size from 1.5 mm to 6 mm (1/16 inch to ¼ inch) in diameter and are usually packaged in drums or coils of up to 1,000 pounds (454 kg) in weight.

Fluxes are granular mineral compounds which are mixed according to various proprietary formulations. They may be fused, bonded, or agglomerateo and sometimes more than one type is mixed for highly critical or proprietary applications.

AWS specifications use a classification system to describe flux-electrode combinations, for example, F7A6-EM12K is a designation for a flux and electrode combination that will produce weld metal with 70 ksi tensile strength in the as-welded condition, with Charpy V-notch impact roughness of 20 ft-lb at −60°F when produced with an EM12K electrode.

OTHER COMMON JOINING AND WELDING PROCESSES

Electroslag Welding (ESW)

The electroslag welding (ESW) process involves a molten slag that melts the filler metal and the surfaces of the work-pieces to be joined. The volume of slag moves along the full cross section of the joint as welding progresses; the slag shields the weld pool. The process starts as the arc is initiated. The arc heats the granulated flux and melts it to form the slag. As a sufficient volume of slag is created, the conductive slag extinguishes the arc. The existing slag is kept molten by its resistance, as the electric current passes between the electrode and the work-pieces.

The ESW process has an extremely high rate of deposition, and it can weld a very thick section of material in a single pass. The process requires minimum joint preparation thus reducing the cost of material handling, which could add significantly to the overall cost of welding and fabrication. Due to the very high deposition rate and use of single pass welding, material distortion is minimal.

The ESW process can only be used on carbon steel and low alloy steels, and must be performed in the vertical position. Welding must be completed in one cycle and cannot be interrupted. The materials to be welded must be at least 19 mm (¾ inch) thick.

Plasma Arc Welding (PAW)

In the plasma arc welding (PAW) process, the heat is produced between an electrode and the work-piece by heating them with a constricted arc. Shielding is obtained from hot ionized gas delivered through a torch. A supplementary shielding gas is usually provided. From the process point of view, it is the constricted gas flow that differentiates the PAW process from the GTAW process.

The plasma issues from the nozzle at about 16,650°C (30,000°F), which allows better directional control of the arc, and the high heat results in deep penetration and a very small heat affected zone. The major disadvantage of PAW is high equipment expense.

As stated in the introduction to this process, the PAW process is about striking an arc, and forcing the hot ionized gas through the nozzle. The PAW process is an extension of the gas tungsten arc welding (GTAW) process. However, it has much higher energy density and greater velocity of the plasma gas through a constricting nozzle.

The process involves directing the orifice gas through the constricting nozzle's plenum chamber. This is formed by the tungsten electrode in the center and the walls of the constricting nozzle. The chamber's exit is constructed to give a tangential vector, which gives a swirl to the exiting plasma gas. The throat length and the orifice diameter define the constricting nozzle. The tungsten electrode located in the center of the constricting nozzle is set offset (set at a distance) from the opening of the constricting nozzle. This distance is called electrode setback distance. This is a significant departure from the GTAW process, where the electrode extends out of the nozzle to strike the arc on the work-piece.

The offset of the tungsten electrode allows for the *collimation* of the arc, which is then focused onto a relatively very small area of the work-piece. Since the shape of the arc is cylindrical in contrast to the focused beam, there is no change in the area of contact as the standoff varies during the welding process. Further, as the orifice gas passes through the plenum chamber, it is heated and expanded, increasing in volume and pressure; this increases the velocity of the gas that exits from the orifice. This is a very important variable in welding, as too powerful a gas jet can cause turbulence in the weld pool. Hence the gas flow rate at the orifice is controlled at 0.25 to 5 lm³/min (0.5 to 10 ft³/hour). An additional shielding gas is introduced to protect the weld pool from atmospheric contamination. The gas flow rate of shielding gas is kept in the range of 10 to 30 lm³/min (20 to 60 ft³/hour).

The effect of the plasma jet created by constricted flow is:

1. Better directional stability, due to its ability to overcome effects of magnetic fields.
2. Higher current density and temperature can be produced.

It may be noted here that the heat produced by an unconstricted arc, like that produced in the GTAW process, is high enough to melt most metals that are welded. Hence the objective is not to generate too high a heat, but to get more directional stability and focusing ability of the plasma jet, and thereby of the plasma arc. This is an efficient use of the energy supplied by the process.

The degree of arc collimation, arc force, and energy density available on the work-piece are functions of several parameters. The following parameters can be altered to produce very high to very low thermal energies, as required by the work-piece:

- Plasma current
- Orifice diameter
- Type of orifice gas
- Flow rate of orifice gas
- Type of shielding gas.

Two different arc modes – transferred arc and non-transferred arc – are used in the plasma arc welding process.

In transferred arc mode, the arc is transferred from electrode to the work-piece because the work-piece is made part of the electrical circuit. Mostly positive polarity is used for welding of all steel and nickel alloys. Light alloys, like aluminum and magnesium, are welded with DCEP or AC with continuous frequency stabilization. In the transferred arc system, the heat is obtained from the anode spot on the work-piece, as well as from the plasma jet. Obviously this system has the advantage of greater energy over the non-transferred arc system.

In contrast, in the non-transferred arc system, the work-piece is not part of the circuit. The arc is established and maintained between the electrode and the orifice of the constricting nozzle. The plasma jet is the sole supplier of the required heat in this system. This process has relatively low energy, and is especially useful for cutting and joining non-conductive materials.

The main advantages and limitations of this process are listed in Table 2-3-4.

STUD WELDING

Stud welding is a general term used to describe joining a metal stud or similar part to a metal work-piece. Welding can be done with many processes, such as arc, resistance, friction, and percussion.

Arc stud welding joins the base (end) of the stud to the work piece by heating the stud and the work with an arc drawn between the two. When the surfaces to be joined are properly heated, they are brought together under low pressure, and the two join at the interface.

TABLE 2-3-4 Advantages and Limitations of PAW Process

	Advantages of PAW process	Limitations of PAW
1	Energy concentration is greater, resulting in: • Higher welding speed • Lower current required to produce given weld • Lower Shrinkages and distortion. Adjusting welding variables can control the depth of penetration. Keyhole technique allows for higher thicknesses being welded with minimum distortions, minimum addition of weld metal, and the wine glass appearance of weld cross section.	Very low tolerance for joint misalignment.
2	Arc stability is improved.	Manual PAW is not very feasible to use, hence automation is the most practical application.
3	Arc column has greater directional stability.	The constricting nozzle requires regular inspection and maintenance.
4	Higher depth-to-width ratio results in less distortion.	
5	Fixturing cost can be reduced.	
6	This operation is much easier for adding filler metal to the weld pool since torch standoff distance is generous, and the electrode cannot touch the filler or the weld pool. This also eliminates tungsten contamination of the weld.	
7	Reasonable variations in torch standoff distance have little effect on bead width or heat concentration at the work, which allows for out of position welding.	

Developed during installation of automatic welding plant at NITIN Castings Limited Bombay; 1984–1985.

Capacitor discharge stud welding is performed with heat derived from the rapid discharge of electrical energy stored in a bank of capacitors. The rest of the process is similar to arc stud welding. As described above, the different types of stud welding processes are similar in all respects except in the manner of heat application to the metal surfaces to be welded together.

Oxy-fuel Gas Welding (OFW)

In the oxy-fuel gas welding (OFW) process, the base metal and filler metal are melted using a flame which is produced at the tip of a welding torch, called the nozzle. In this process, a fuel gas and oxygen are combined in a mixing chamber and ignited at the nozzle tip.

The process derives the heat for welding by burning two gases; sometimes the process uses part of the required oxygen gas from the supply and supplements the rest from the atmosphere, which is the case for the oxy-acetylene process.

One advantage of OFW is the independent control that the welder has over the heat and the filler metal. This gives the process a clear advantage for use in repair welding, and for welding thin sheet and tubing.

The equipment for most of the oxy-fuel process is low cost, portable, and versatile.

The process is also used as a cutting tool. Cutting attachments, multiflame heating nozzles, and other accessories are available. Mechanized cutting operations are easily set up and widely used in a variety of ways.

For cutting, the flame is used to heat a spot on the metal to be cut. When the required temperature is reached, extra flow of oxygen is introduced to burn off the metal, a continued heating and flow of oxygen effect cutting.

Brazing and Soldering

Brazing is the process in which the base metal is not melted. The process joins materials by heating them in the presence of filler metal, which melts and spreads on the abutting surfaces of the metals by wetting and capillary actions. The liquidus temperature of the filler metal is above 450°C (840°F) but below the solidus of the base metal.

Soldering follows the same principles as brazing, except that the filler metal liquidus is below 450°C (840°F). The base metal liquidus is much higher than that of the filler metal. The filler metal distributes itself between the closely fitted surfaces of the joint by capillary action.

Parts must be properly cleaned and protected, either by flux or a controlled atmosphere, during the heating process to prevent excessive oxidation. The filler must be provided with a capillary opportunity, and the heating process must provide the proper brazing temperature and a uniform heat distribution.

Heat sources include furnaces, induction heating coils, resistance heating, dip brazing in molten metal or molten chemical baths, and infrared brazing.

AWS Specification A5.8 divides filler metals into seven different categories, and subdivides these into various classifications within the categories. Aluminum-silicon, copper, copper-zinc, copper-phosphorus, silver, gold, nickel, and cobalt fillers are commonly available.

Joint clearance has a major effect on the mechanical performance of a brazed joint. Clearance controls the mechanical effect of restraint to plastic flow of the filler metal, and too much leads to defects like slag entrapment, or voids. Capillary force also affects the amount of filler that must be diffused with the base metal.

Fluxes are often required to react with the oxidized surfaces of the base metal to clean them and so maximize the flow of the filler metal. Fluxing also roughens the surface to enhance wetting and capillary flow of filler metal.

ARC-WELDING POWER SOURCES

As we have seen in the discussion above, there are various arc-welding processes, and many of them also have a number of sub-variants. All these processes use specialized power sources that provide specific characteristics of the electric arc to give that special edge to the process. We shall discuss some of the more common power sources in the following paragraphs.

The available line voltage in industry is generally very high. However, welding requires low voltages; hence the line voltage is not fit for welding use in the as-supplied condition. The welding power source must be able to reduce this high input voltage of 120V, 240V or 480V to a suitable output voltage, which is usually in the range from 20 to 80V. Transformer, solid-state inverters or motor generators are intermediary components that step down the line voltage. They are all capable of converting the line voltage to an input voltage suitable for welding.

Some welding sources are linked with their own prime mover that is linked to a generator or alternator that produces the required open circuit voltage for arc welding.

As these intermediary components convert high voltage to low voltage they also provide high welding current in a general range of 25 to 1,500 amperes (A). The output of a power source can be either an alternating current (AC) or direct current (DC), although some sources are capable of producing both AC and DC, and the required current can be obtained by changing the switch. The output may be either constant-current, or constant-voltage or both. Some power sources have a mode that can provide pulsating output. Transformers can deliver only AC current, but a transformer coupled with rectifier system can deliver both AC and DC current. Because of this type of cross variation, the sources are often described in very different ways; for example, a power source for GTAW process could be called, 'a constant-current AC/DC transformer rectifier'. These sources are further labeled with their current rating, input power requirements, duty cycle etc. Further classification would include such special features as current pulsing ability, line voltage compensation, remote controlled, water or air-cooled, high frequency stabilization etc.

The classification of constant-current or constant-voltage sources are based on the static volt-ampere characteristics. For conventional power sources the terms are very relative, as in fact there is hardly any output that is actually constant-voltage or constant-ampere. However new equipment is becoming available that is capable of holding the current or voltage truly constant. Constant-voltage sources are also referred to as constant potential power sources and similarly, constant-current sources are also called variable voltage units.

Constant Voltage Power Source

Constant voltage power sources are used in welding processes that are self-regulating and use a constant feed electrode, mostly in the form of wire. The system stabilizes the arc with the constant changes of the torch position. The arc current is kept in a nearly fixed proportion to the wire feed rate. The National Electrical Manufacturers Association (NEMA) in its publication Electric Arc Welding Power Sources, defines a constant voltage source as 'a constant current AC/DC transformer rectifier'.

Figure 2-3-3 shows a typical volt-ampere output relationship for a constant voltage power source. Due to the internal electrical impedance there is a minor drop in the slope. Note that from point Y on the slope, corresponding to 200 amps on the X-axis, any variation in voltage (on Y axis) to points X or Z, the slope has a corresponding change in the current. The volt-ampere characteristics shown by this curve are suitable for any of the constant electrode feed welding processes, including GMAW, SAW and FCAW processes. It is very well known that for these welding processes, a slight change in arc length (voltage) causes a large change in the welding current.

The slope also explains the difference between static and dynamic power supply characteristics. In the short circuit GMAW process, for example, when the electrode tip touches the weld-pool the electric circuit is shorted, and arc voltage approaches zero. At this point, it is the inductance that limits the rapid increase in welding current. The dynamic characteristics designed into the power source compensate for the action by rate of current change, and prevent any explosive dispelling of the molten weld metal.

Constant-Current Power Source

Due to the variation in arc length, manual welding processes use this type of power source, including both SMAW and GTAW. However with voltage sensing devices the power source can monitor and correct the arc length, meaning that these power sources can be used for semi-automatic and automatic welding processes like GMAW, FCAW and GTAW.

NEMA defines a constant-current arc welding machine as a 'constant current AC/DC transformer rectifier'.

In this type of system, each current setting results in a separate volt-ampere curve if tested with a resistive load. The open circuit voltage (OCV) of this type of source is significantly higher than the arc voltage. Figure 2-3-6 describes the typical 'drooping' arc characteristics of a power source with adjustable OCV. Such a power source may have both OCV adjustment and output current control. A change in any of the controls will change the slope of the volt-ampere curve. In Figure 2-3-4, curve A represents an OCV of 80 volts. Here a steady increase of arc voltage by 25%, from 20 to 25 volts, results in a relatively small current decrease of 6.5%; from 123 A to 115 A. Under this condition, for

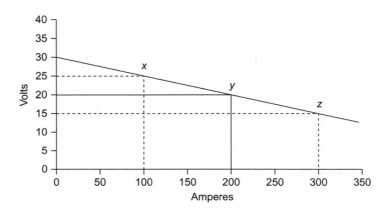

FIGURE 2-3-3   Volt-ampere output relationship for constant voltage power source.

consumable electrode welding processes such as SMAW the electrode-melting rate would remain fairly constant with very slight changes in arc length. However, if the OCV of the power source was set lower – for example at 50 V – then the same 25% change in arc voltage from 20V to 25V would cause a significantly higher – about 19% – change in current; from 123 A to 100 A. For manual welding processes, an experienced welder would prefer a flatter Volt/Ampere curve power source, because this would give him the option of varying the arc length for an out-of-position welding, as in pipeline welding.

Transformers

Welding transformers are simple transformers that have primary and secondary winding coils. The secondary coil has taps to vary the length of the coil and to draw the selected current output. This also controls the OCV. The transformers work on the principle of maintaining a fixed ratio of input to output. There is a fixed relationship between input and output voltage, and an inverse relationship between input and output current, as given below:

$$N_1/N_2 = E_1/E_2 = I_2/I_1 \tag{1}$$

Transformers are designed by inserting an impedance circuit in series with the secondary winding to provide output volt-ampere slope characteristics that can be adjusted for welding conditions by selecting a specific tap.

In constant-current power units, the voltage drop across the impedance (E_x) increases as the load current is increased. This change in voltage drop causes a significant reduction in arc voltage (E_A). Adjustment to the value of the series impedance controls the voltage drop and the relationship between load current and load voltage, this is called 'slope control' or sometimes 'current control'.

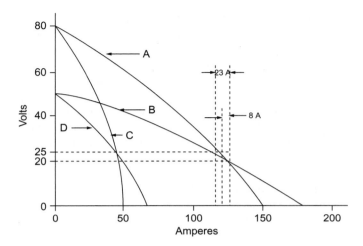

FIGURE 2-3-4   Volt-Ampere characteristics of a drooping slope power source with adjustable OCV.

In a constant voltage source of power, the output voltage is very close to the requirement of the arc. In this case, E_x across the impedance reactor increases only slightly as the load current increases, the reduction in load voltage being small. Control over the relationship between load current and load voltage is achieved by adjustment of value of reactance, though it may be noted that this control is only very limited.

Thyristor-Silicon Controlled Rectifiers (SCR)

The term solid-state device derives from the science of crystalline solids. Unlike rectifiers that convert AC to DC, silicon diodes work as one-way valves when placed in the circuit. A proper arrangement allows conversion from AC to DC.

Solid-state devices have replaced the shunts, reactors, moving coils etc. that were previously used to control the output of a welding transformer, since they can control welding power directly by altering the current or voltage wave form. A thyristor – a type of silicon-controlled rectifier (SCR) – is one such device used for this purpose.

The two main functions of SCR are in phase control mode with transformers, and in converter configuration.

The most common SCR phase-controlled machines are three-phase machines in either constant current or constant voltage mode. Because of the electronic control of output characteristics, automatic line voltage compensation is easily achieved; this allows the machine to set the precise welding power, independently of the variations in input power. An SCR can also serve as a secondary contactor, that allows the welding current to flow only when the control allows the SCRs to conduct. This is a very useful feature in spot and tack welding operations, where rapid cycle operation is required. An SCR contactor does not provide electrical isolation; this neeeds a circuit breaker or similar device to be provided for electrical safety.

A transistor is another solid-state device, but unlike SCRs they are expensive and provide a very precise number of variables to the power supply, thus limiting their use in common application machines. Conduction through the device is proportional to the control signal applied. If a small signal is applied there is a corresponding small increase in conduction, and if there is no signal there will be no conduction. Unlike SCRs the control can turn off the device without waiting for polarity reversal or 'off' time. The disadvantage is that transistors do not carry current like SCRs, which means that several transistors are required to match the output of an SCR.

Transistors are used in power sources to modulate frequencies, pulse widths etc.

Generators

In arc welding, rotating equipment is often used as a power source. These are generators that produce direct current (DC) and alternators that produce alternating current (AC).

Main or shunt field windings are used to control the no-load output of a generator. A filter reactor or an inductor is replaced with several turns of series winding on field poles of the rotating generator to provide the inductance that is required for a stable arc. These machines allow the polarity to be changed by interconnection between the exterior and the main field.

Alternators

An alternator is a rotating type power source that produces alternating current, which may be used as it is, or rectified through a rectifying circuit to be used as DC current for welding. For rectifying it may use any of the methods discussed in this chapter.

Chapter 4

Physical Effect of Heat on Material During Welding

In previous chapters we have discussed and understood that in all welding processes we use heat to make the joint. Apart from a few exceptions, all welding processes use the direct application of heat for welding, though the method of making that heat varies from process to process.

This application of heat is localized near the weld joint and affects the parent metal closer to the weld joint. This autogenous heating for welding is often performed by the current flow and contact resistance of the two metals being joined. In the history of the development of welding, almost all possible heat sources have been used. Some of the sources of heat in welding are indirect byproducts of a primary process like friction, ultrasonic or explosion methods.

Irrespective of how the heat is applied in a specific welding process, it is essential that the process is in control of the conditions of heating. The thermal conditions are responsible for the changes in metallurgical structure, mechanical properties, and residual stresses and distortions created in the material. The four factors that are influenced by heat and that affect the material properties are:

1. The distribution of maximum temperature, also called the 'peak temperature'.
2. Distribution of heat between the weld-metal and the heat affected zone (HAZ).
3. Cooling rate in the weld-metal and heat affected zone (HAZ).
4. The solidification rate of the weld-metal.

The chapter on the physics of welding discussed how to model the effect of heat conditions in the weld. In this chapter we shall discuss the practical

171

applications of these effects of heat. For this discussion we will consider the effect of heat in a simple 0.20% carbon steel weld.

THE MOLTEN METAL

As soon as the flame or arc is directed onto the steel, its temperature rises. The heat absorbed from the heat source spreads beyond the spot directly underneath. The temperature rises steadily at first. At about 510°C (950°F), any distorted grains resulting from previously working the steel start to recrystallize.

The steel immediately surrounding the weld metal is in a mushy condition – there is a mixture of crystals and liquid. Inside the 850°C (1,560°F) curve, the steel is in the gamma (γ) form, while between 510°C (1,560°F) and 725°C (1,335°F) the steel is a mixture of alpha (ferrite) and gamma (austenite) crystals.

Above 850°C (1,560°F), the austenite crystals grow as many small austenite crystals coalesce to form fewer larger crystals.

At 1,490°C (2,714°F), melting begins, and continues up to 1,520°C (2,768°F) when the last of the crystals melt.

In the reverse cycle, as freezing begins at 1,520°C (2,768°F), crystals of austenite begin to appear. As cooling continues, the proportion of austenite crystals present increases and the melt becomes mushy.

Tiny crystals start to appear randomly in the liquid. The crystals grow in preferred directions, until they come into contact with other growing crystals to form a polycrystalline metal. The initial crystals growing in the weld metal are columnar and the later crystals are multiaxial. Between 1,490°C (2,714°F) and 850°C (1,560°F), there is no change in the austenite crystals.

At 850°C (1,560°F), ferrite crystals separate out from the austenite at the austenite grain boundaries. Just above 725°C (1,335°F), 25% of the steel is ferrite and 75% is austenite.

Cooling continues until at 725°C (1,335°F) the austenite starts to transform to pearlite.

The temperature distribution in a plate is graphically described in Figures 2-4-1 and 2-4-2.

THE WELDED PLATE

All parts of the plate that attained a temperature of about 1,100°C (2,000°F) will have the structures described above since the austenite transformations will have occurred.

The portion of the plate that was heated to above 850°C (1,560°F) and up to 1,100°C (2,000°F) will have a more ductile structure. The faster cooling rates in this heat range produce very small austenite grains. The ferrite grains in the plate material would be distributed throughout within the pearlite grains.

In the plate-area that reaches 725°C (1,335°F) to 850°C (1,560°F), the pearlite grains would transform to austenite, and some of the ferrite will be absorbed

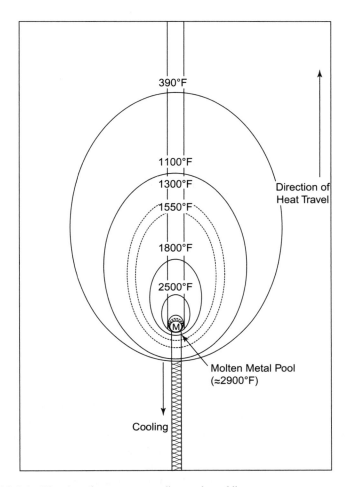

FIGURE 2-4-1   Direction of temperature gradient as the welding progresses.

into the austenite grains. During cooling, small ferrite grains separate and at 725°C (1,335°F), the remaining austenite transforms back to pearlite. The final structure consists of coarse ferrite grains which never dissolved, and much finer clusters of ferrite and pearlite grains where previously only pearlite existed.

INFLUENCE OF COOLING RATE

The faster the weld metal is cooled, the greater tendency it has to undercool and the grain size of the solidified weld is smaller. Faster cooling rates also favor the formation of trapped slag inclusions and gas blowholes. Because of the allotropic changes that occur in steel, the cooling rate at temperatures below 850°C (1,560°F) influences the structure.

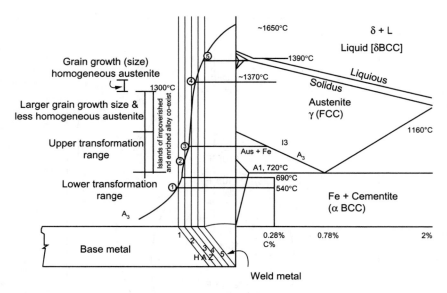

FIGURE 2-4-2   Effect of heat on weld, superimposed on phase diagram to show metallurgical changes.

Very rapid cooling from 850°C (1,560°F) to 750°C (1,380°F) causes the austenite to undercool rather than transform immediately to pearlite. Because there is insufficient time for the diffusion of carbon atoms required for the austenite to pearlite transformation, a massive transformation to martensite occurs instead, at about 315°C (600°F). The carbon atoms are trapped in the distorted body-centered tetragonal martensite crystals.

In Figure 2-4-1, the distribution of heat in a plate is described by lines that indicate the typical pattern and distribution that occurs.

The same effect is shown in Figure 2-4-2 for a weld. In this picture, the heat distribution in the weldments is superimposed on an iron-carbon phase diagram indicating the possible effect of heat on various regions of the weldments. Such superimposing is helpful in understanding the effect of heat.

Chapter 5

Stresses, Shrinkage and Distortion in Weldments

Because of the unavoidable effects of heat that always accompany welding, dimensional changes will occur. However, they can be minimized and often one condition can be used to counteract another.

Weld metal shrinks upon solidification, but this is not the primary cause of distortion problems in weldments after welding. During solidification, as the atoms of iron in the melt assume their fixed positions in the crystal lattice of growing solid grains, the coupling of the liquid and solid are very weak. This means that the weld metal cannot exert much stress on the adjacent base metal. Solidification shrinkage accounts for dishing or deformation in the weld metal.

Applied Welding Engineering: Processes, Codes and Standards.

It cannot, however, generate stresses capable of decreasing the overall size of the weldments or pulling a portion of it.

Immediately following solidification, however, the cooling weld metal continues to contract. This thermal contraction generates stresses up to the yield strength of the material at that temperature in the cooling cycle. This could lead to some very serious damage to the material, including possible cracks.

Distortion is deviation from the desired form and occurs in welding because of stresses which develop in the weldments from localized thermal expansion and contraction. Distortion is dependent on the following:

1. The magnitude of the welding stresses developed by localized thermal expansion and contraction.
2. The distribution of these stresses in the weldments.
3. The strength of the members.

STRESSES IN WELDMENTS

Definitions of Terms

Residual Stress

Residual stress is the internal stress that remains in a member of a weldment after a joining operation. They are generated by localized, partial yielding during the thermal cycle of welding, and the hindered contraction of these areas during cooling.

Structure Stress

Structure stress arises from grain boundaries, crystal orientations, and phase transformations in small volumes of weld metal.

Reaction Stress

Reaction stress is an internal stress which exists because the members are not free to move.

Stress Concentration

Stress concentration refers to the increased level of applied stress which develops at abrupt changes in section such as sharp transitions, abrupt changes in weld profile, sharp corners, notches, and cracks.

DEVELOPMENT OF STRESSES

Moving a Localized Heat Source

The local rise and fall in temperature at any point along the weld, as the heat source first advances toward a point and then passes it, develops stresses and causes changes in the microstructure.

If the metals being welded had zero coefficients of thermal expansion, no stress would develop and shrinkage and distortion would not occur. But temperature changes do cause a change in volume, and since in welding the parts are never free to expand or contract in all directions, it is the behavior of steel attempting to expand and contract under conditions of restraint that must be considered.

Nearly all welds are started at room temperature and regardless of the restraint, the maximum tensile shrinkage stress is close to the yield strength.

DISTRIBUTION OF STRESS IN A SIMPLE WELD

The sketch below describes the stresses that build in weldments. With the help of Figures 2-5-1 and 2-5-2 we can understand the effect of heating and cooling

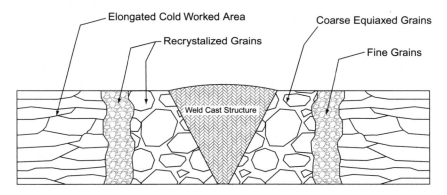

FIGURE 2-5-1   Cause of stresses in Weldments.

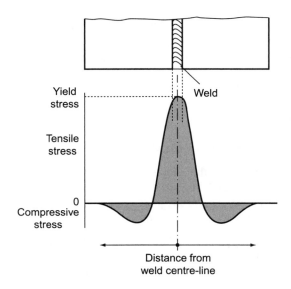

FIGURE 2-5-2   Stress distribution in weld.

in a simple weld. In more complex joints, the stresses are more complex and have more serious implications on the performance of the weldment.

A weld is rapidly deposited along the edge of two pieces of metal. The entire weld zone is still at a high temperature when the weld is completed.

At high temperature, the metal close to the weld attempts to expand in all directions, but it is prevented (restrained) by the adjacent cold metal.

Because it is being restrained from lateral expansion (elongating), the metal close to the weld is compressed.

During cooling, the unstable zone attempts to contract, once again, now the cold metal surrounding it restrains it. As a result, the unstable zone becomes stressed in tension.

When the welded joint has cooled to room temperature, the weld and the unstable region close to it are under residual tensile stresses close to the yield strength.

To balance the shrinkage-related tensile stresses developed at the edge, there must be a region of tensile shrinkage stresses opposite in the unwelded edge and a region of compressive stresses between the two tensile zones.

RESIDUAL STRESSES

In order to calculate residual stresses in weldments, detailed, accurate information of the temperature distribution during the welding process is needed. Variations in thermal conductivity and changes in mechanical properties at high temperatures further complicate the calculation.

It is known that stress gradients similar to temperature gradients exist in the metal adjacent to a weld. Restraint during welding, such as rigid clamping, or even the mass of the structure itself, results in even higher residual stresses. Since the stress system is in equilibrium, externally applied loads cannot add to the residual tensile stresses until the residual compressive stresses are overcome.

If the load does cause a small amount of plastic strain in highly stressed areas, then the peak stresses in those areas will be reduced. Stress relief heat treatment may dissipate residual stresses.

SHRINKAGES

Shrinkage Transverse to a Butt Weld

Shrinkage perpendicular to the long axis of a weld is called transverse shrinkage. It is primarily dependent on the cross-section of the weld metal in the joint, as well as the thickness of the joint.

The amount of shrinkage varies with:

1. Degree of restraint on the members during welding and cooling.
2. Cross-sectional area of the weld metal.
3. Extent to which it flows into the adjacent base metal.

4. The number of beads or layers in the weld.
5. Temperature-time cycle of the bead deposition.

Transverse shrinkage is cumulative. For welds made in 25 mm (1 inch) and thicker steel by the SMAW process, an estimated calculation of transverse shrinkage is made using the following relationship between base metal thickness, cross-sectional area and the root opening of the weld:

$$\text{Transverse shrinkage (S)} = 0.2\,(A_w/t) + 0.05\,(d) \tag{1}$$

Where:

S = Shrinkage liner (inches)
A_w = The cross-sectional area of the weld
t = The base metal thickness, and
d = The root opening (US customary units).

For a calculation in SI units, the following equation can be used:

$$S = 5.16 \times A_w/t + 1.27\,d$$

In general, a greater degree of restraint results in less shrinkage and higher residual stress levels.

In practice, it has been observed that welding in flat or horizontal positions allows higher travel speed and decreases shrinkage by up to 20% compared to other out-of-position welds.

Preheating provides more uniform heating and cooling, and minimizes shrinkage.

Cooling between passes increases restraint and reduces shrinkage.

Although post-heating reduces residual stresses and promotes plastic flow, there is no change in the shrinkage which occurred during welding.

Cold peening can counteract transverse shrinkage, however the undesirable side effects of peening must be considered before specifying it as a control method for transverse shrinkages.

SHRINKAGE LONGITUDINAL TO A BUTT WELD

Longitudinal shrinkage is proportional to the length of the weld. It is also a function of the weld cross-section and the cross-section of the surrounding, colder, base metal, which resists the expansion and contraction forces of the heated weld and base metal.

A rough estimation can be made using the formula for predicting longitudinal shrinkage when the restraining plate is not more than 20 times the cross-sectional area of the total weld:

$$\text{Longitudinal shrinkage (S)} = 0.025(A_w/A_p) \tag{2}$$

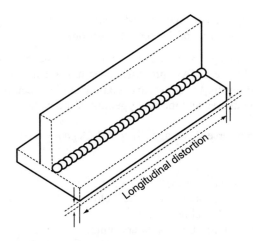

FIGURE 2-5-3   Longitudinal stress in weld.

Where:

S = Shrinkage in linear inches
A_w = The area of weld, and
A_p = The area of the restraining plates.

In another approach to the calculation of longitudinal shrinkage (using values in SI units) the following equation is used:

$$\Delta L/L = 3.17 \times I \times L/100,000 \times t$$

Where:

L = Longitudinal shrinkage (mm)
L = Length of weld (mm)
t = Thickness of the plate (mm)
I = Welding current (Amps).

DISTORTION IN WELDMENTS

General Description

The localized area along which the arc or heat source passes is the starting point of a distortion problem. The temperature differential between the weld zone and the unaffected base metal is great, and much localized expansion and plastic flow take place. Restraint from clamping and the mass influences the extent of plastic flow.

Angular Distortion

Angular distortion is the angular change in relative positions of members extending from a weld area, as shown in Figure 2-5-1. Note that there is a shorter width of contracting weld metal at the root of each weld than there is at the face. This difference in width, which must contract upon cooling, is a major factor in angular distortion.

For single-bevel groove welds in butt joints and T-sections, the amount of angular distortion is nearly proportional to the number of beads or layers deposited in a joint.

Angular distortion can be controlled by the following:

- Using the minimum amount of weld metal required
- Depositing the weld metal in the fewest number of layers
- Avoiding as much as possible very narrow root profiles and wide faces
- Balancing the amount of weld metal about the neutral axis of the weld
- Presetting the members at a slight angle opposite the location where distortion is expected to develop.

Longitudinal Bowing

Longitudinal distortion of long members is caused by shrinkage stress which develops at some distance from the neutral axis of the member. The amount of bowing is determined by the magnitude of the shrinkage stress and the resistance of the member to bending, as indicated by its moment of inertia.

Bowing on a long slim member can be roughly calculated in U.S. customary units as:

$$\Delta = 0.005\, A_w\, (L^2)d/l \qquad (3)$$

Where:

Δ = The resulting vertical movement
A_w = The total cross-sectional area within the fusion line of the welds
d = The distance between the center of gravity of the weld group and the neutral axis of the member
L = The length of the member (inch), and
l = The moment of inertia of the member.

Longitudinal distortion is controlled by balancing welds around the neutral axis of the member. Prebending the member in a bow in the opposite direction to that which will develop from welding is often practical.

Buckling

Thin sheet is subject to buckling, which is caused by the inability of a laterally unsupported sheet to resist compressive stress without buckling.

Welding alternately on opposite sides of a part will reduce distortion by neutralizing each distortion.

CORRECTIVE MEASURES

Corrective measures for distortions of any of the types discussed above must be pre-assessed, and weldments designed accordingly. Failure to acknowledge distortions and provide distortion control in design often results in very difficult situations during and after the fabrication is completed, possibly leading to repairs or even rejections.

Appropriate controls include activities during execution, planning and sequencing of welds including the pre-heat, post-heat treatments, counter allowances for distortions, staggering of the welds, choice of the welding processes, control of artificial restrains etc. Post-fabrication correction is always difficult and expensive, and sometimes disastrous for all the efforts and costs incurred in fabrication.

Marginal corrections can be accomplished on some sections of the weldments. The success of the effort is dependent on how restrained and strained is the member that is being corrected.

Application of heat to achieve the correction is one safe method, and is often practiced with somewhat limited success. The result of such straightening is dependent on various points discussed above, including the inbuilt restraint of the member itself.

THERMAL STRAIGHTENING

Thermal straightening is the deliberate application and controlled shrinking of specific portions of plates or structural members. This is a controlled application of the problem itself to correct the problem, in which a localized area of metal is torch-heated to a high temperature – for most structural steel this is 850°C to1000°C. During heating, the localized area must expand, but it is constrained by the colder, stronger surrounding metal. So the heated metal is forced to upset or bulge in the desired direction.

When the localized area reaches a reasonable red heat (870°C to 990°C, 1,600°F to 1,800°F), heating is stopped to confine the upsetting to the desired area and direction.

Upon cooling, the heated area contracts, but because the upset is not equally reversible during cooling, a net shrinkage results which pulls the surrounding metal inward. So the movement in this direction will be greater than the movement in the opposite unwanted direction during heating.

This movement is used to control the distortion that occurred earlier during welding or by some mechanical action.

Buckling in a sheet can be removed by heating a localized area in the center. The surrounding metal must be kept cool and preferably restrained. Upon cooling, the central localized spot will shrink and draw the sheet flat.

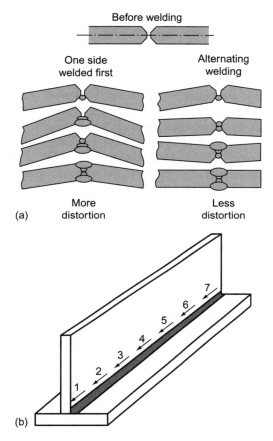

FIGURE 2-5-4   Corrective measures (a) controlled distortion of weld. (b) Controlled welding technique.

DESIGNING WELD JOINTS

The design of a weld is part of the overall activity of designing the structure. Thanks to the popular television series, everyone understands that a structure is considered to be only as strong as its weakest link. In a welded structure, or 'weldment', the weakest link is often the weld joint. This is because the weld has a different metallurgical structure to the parent metal. Often, attempts are made to overcome this by either over-designing the weld or by over-matched weld metal.

In the following paragraphs we will briefly discuss some of the factors that need to be considered when designing a successful weld.

Earlier in this chapter we discussed stress in welds, and the focus was on the stresses caused in metal due to the heating and cooling cycles of welding, and resultant effect on the shape and appearance of the weldments. In this part we concentrate on the determination of some numerical values of stress, and matching those values with the overall structural requirements.

Assessing the Strength of Welds

There are a number of factors to consider when designing a structure; one of them is the allowable stress (f) in its members. When a structural member is constructed or joined to another member by welding, it is imperative that strength of the weld is also determined as a consideration for the integrity of that structure. This means that a welding engineer will need to determine, among several other things, the type of material, the available welding processes and their advantages and limitations, and also what type of weld joint is best suited for the purpose. The options are numerous – for example, to use full penetration welds that may involve V, J, U or combinations of weld groove preparations, or a lap or a fillet weld, whether to use continuous weld or skip or stitch weld etc. This decision will be further affected by the fact that the weld joint may or may not be accessible from both sides. For example the girth welds in pipeline welding are seldom accessible from the inside; hence a full penetration weld from one side is made. However, for similar girth welds in a pressure vessel shell, it is usually possible to back gouge and weld from the inside. To understand all these issues, the terminology relating to weld design must be understood; hence some of the terms used in designing the welds are explained below.

Throat of a Weld

The throat of a weld is its effective depth. In a full penetration weld it will be the depth of the groove measured normal to its base. In a fillet weld it will be measured at 45° from the root of the fillet to the hypotenuse of the weld. In a right angle weld (a weld with equal leg length) the 45° will intersect the hypotenuse in the middle, however this will not intersect in mid-hypotenuse if the leg lengths are unequal. The throat is often denoted as (t_e) meaning effective throat.

Like any structure, for welding the stresses that need to be considered are tension, compression and transverse stress (shear). For a butt weld, if the welding consumable that is selected for the welding would result in a weld metal that will have tensile strength and ductility that matches the parent metal, then the weld is considered suitable for the tensile and compressive load services of the structure.

For fillet welds or members that would be in shear stress, the determination of allowable stress (f) is different from butt welds and its calculation will now be explained.

The allowable stress for a fillet or partially penetrated weld is denoted by Greek letter τ, and it is established as equivalent to 30% of the specified tensile strength of the welding consumable used for welding. Using this information the calculation of unit force per linear inch for a given size of the weld for a given weld metal can be determined.

If we assume that a given equal leg (ω) fillet weld is of 3/16" (0.188-inch) and the weld metal specified tensile strength is 60,000 psi (E60XX welding electrode) then the allowable shear stress for this fillet weld can be determined as follows:

$$f = 0.707\,\omega\,\tau$$
$$= 0.707 \times (0.188 \times 0.30) \times 60\,000$$
$$= 2\,384 \text{ lb/linear inch of the weld} \qquad (4)$$

or

2.38 kips/linear inch of the weld.

The above calculation is a simple example. The mechanical strength calculations for all welds are similar, but the other factors associated with welding will vary significantly. For example, if the weld is in a high strength alloy, then the impact of pre-heat and post-heat treatment of the weld will have to be considered. This fact highlights the importance of the proper welding procedure specification, its qualification to establish the mechanical properties and qualification of welders and welding operators to be able to strictly comply with the specified parameters so that consistent quality of weld is produced.

Sizing a Fillet Weld

The above calculation can also be used to determine the required size of a fillet weld to meet the loading stress. If the allowable stress of the structure is known, then it is relatively easy to determine the size of the weld using the above relationship between various factors.

The selection of welding consumables of matching strength is an essential factor in determining the size of the weld.

Stress Causing Fatigue in Weld

The performance of weldments under cyclic conditions is an important consideration in design and fabrication of structural members. Several examples of fatigue can be cited from common experience; bridges over rivers carrying cars and trucks, overpasses on highways, railroad bridges carrying freight trains and passengers. The fatigues caused in industrial equipment that go through several heating and cooling and pressure cycles also go through fatigue stress in metals and welds of these equipment. Fatigue failure at normal working stresses is generally associated with stress concentration. The fatigue stress is a major cause for concern in offshore deep-water structures and risers. The welding procedures qualified for these applications are tested to meet these stresses.

Welding processes and procedures are carefully selected to avoid any undesirable and abrupt changes in the weld metal. The factors to consider in such an evaluation are not only the match in weld metal tensile strength, or that the weld meets the specified quality, but it must also be ensured that the weld is sound enough to address the following important issues during the expected life of the structure:

- Full penetration of weld
- Welds with no stress risers (e.g. undercuts, notches, lack of fusion, excess penetration, abrupt transition in weld profile, the angle at which the weld profile merges with the parent metal etc.)
- Weld metal ductility, hardness control both in weld metal and in HAZ
- Weld metal yield to strength ratio, in relation with parent metal YS/TS ratio
- Crack propagation
- Crack arrest properties
- Acceptable level of defects and their interaction under stress
- The service environment and its long-term effect on properties of the metal, especially on hardness of base metal as well as weldments.

For a given material, the calculation of maximum allowable stress-causing fatigue in a given life period is dependent on the ratio between minimum and maximum stress (K).

This relationship is expressed as:

$$\sigma \max = \sigma_{ST}/1 - K \tag{5}$$

Where K is the ratio of minimum to maximum stress, and σ_{ST} is the stress in a steady state.

The allowable stress should not exceed the steady stress for the parent material, however, the stresses that cause fatigue failure are more complex and cannot be calculated by the above equation. This complexity is explained by the variables that will now be discussed.

The effect of the weld profile contact angle with the parent metal is plotted on the graph shown in Figure 2-5-5 below, where the contact angle is plotted on the X-axis and the fatigue cycle (2×10^6 in MPa (ksi)) is shown on the Y-axis. Note that a plain plate has a much better resistance to fatigue than a welded joint, and that stresses continue to rise with increasing contact angle. It may be empirically stated that a contact angle of less than 20° gives better fatigue resistance in weldments.

It may be noted that planner defects in the weld increase the possibility of fatigue failures whereas rounded indications like porosity and inclusions are not likely make a major contribution to fatigue crack propagation. In fact it has been shown in several experiments that porosity is able to resist and control the growth or propagating cracks. In this context it is important to note that inclusions which are longer in length and closer to the material surface are to be considered as contributing to the crack growth as compared with rounded inclusion. The depth of these discontinuities is also a factor in their contribution or resistance the growth of fatigue cracks. The deeper defects are in fact compressed under the stress, and often reduce in size. This means to that they are not able to help in the propagation of cracks.

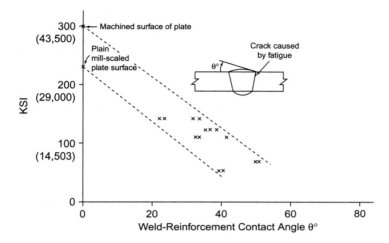

FIGURE 2-5-5   Effect of weld reinforcement contact angle on fatigue strength of weldments.

Fracture mechanics helps determine the relationship between the stress and the critical discontinuity size. The critical discontinuity size is the size of the imperfection that would propagate under plain strain conditions. It is well known that the critical discontinuity size is inversely proportional to the square of the applied stress ($S = 1/\sigma_{applied}$). The geometry and location of the discontinuity assumes critical importance when determining the relationship between the stress and the size of discontinuity. In this process the depth to length ratio plays an important role, and a valid plain strain or elastic-plastic fracture value is required to determine the combination of stress and discontinuity size at which a structure can be safely operated.

The existence of plain strain at the tip of the discontinuity is essential for a discontinuity to initiate a crack. Thus cracks, incomplete fusion or lack of penetration are the key discontinuities that contribute to brittle failure. Porosity and slag inclusions are generally not contributors to such failures. Fracture mechanics allows us to analyze the critical crack size for unstable crack growth leading to fracture. Crack tip opening displacement is one of the common methods for determining the growth of a crack that leads to fracture. In this process, the tip of a sharp crack is loaded under plain strain conditions. If the stress intensity factor, denoted by the letter K, exceeds the critical stress intensity factor K_{IC}, then the crack becomes unstable and propagates, leading to the failure of the member. The relationship of the stress (σ), the stress intensity factor K, and the crack length (a) are described by the following equation:

$$K_1 = C\sigma(\pi a)^{1/2} \tag{6}$$

Where:

C = a constant depending on discontinuity size and shape.

In the engineering critical evaluation (ECA) of weld for fitness for service evaluation and to determine the acceptable size and shape of a discontinuity, some assumptions based on empirical experiences are made; these are listed and explained below:

- The plain strain fracture toughness of the material, K_{IC}, is pre-determined
- The critical size for an unstable fracture is determined from the above equation
- The worst case is assumed; that is the weld discontinuity is a crack
- The vector sum of the applied and residual stress is estimated
- A margin of safety is applied and a maximum allowable crack-like flaw is selected
- Acceptance criteria for more innocuous discontinuities are defined.

Fatigue tests are carried out to determine the behavior of material, and, from a welding engineer's perspective, the behavior of flaws in any weld under stresses causing fatigue. Several tests are involved in determining fatigue behavior. The process may involve introducing stresses to develop fatigue in the member and then determining the growth of the weld flaws, in which case both buried and surface opening flaws are tested. A number of the mechanical tests discussed earlier in the section are involved, including hardness and fracture toughness testing of welds. In welds the method is used to determine the opening mode plane strain fracture toughness (K_{IC}), the critical crack tip opening displacement (CTOD) fracture toughness and critical J fracture toughness of the material. The weld is pre-cracked at the specific location and tested in displacement control under quasi-static loading and at a constant rate of increase in stress intensity factor, which is within the range of $0.5\,MPa.m^{05}_{s}{}^{-1}$ to $3.0\,MPa.m^{05}_{s}{}^{-1}$ during the initial elastic deformation. The method is useful in determining ductile to brittle fracture behavior under decreasing temperature.

In the fatigue testing of pipes, often for offshore deep water steel catenary risers (SCRs), a full scale fatigue testing method using resonance testing machines is adopted, in which a rotating bending moment of constant amplitude is applied at the resonant frequency of the pipe (typically between 25 Hz to 30 Hz) allowing the high-cycle testing of riser girth welds. The system is used for up to 200 million cycles.

Weld Size and Cost Control

Welding cost analysis involves taking account of the weld design, welding process selection, and preparation, labor and material. Several commercial software packages are available to calculate the cost of welding. They take all the data, and produce a cost estimate. However, the most visible and well-known factor – though not the most expensive – is the dependence on various welding consumables, particularly welding filler metals, in the form of filler wires or electrodes. The control of the cost of electrode filler wire can make a significant difference to the profit or loss of a project. Another side effect of over-welding is not so much the cost of consumables, but the possible stresses that this introduces, which may lead to premature failure of the weldments, and metallurgical changes in material affecting the performance of the weldments.

One of the main factors to consider in designing a weldment is the cross-sectional area of the weld. The weld cost is directly related to the cross section of the deposited weld metal. However, while controlling the weld metal volume, it must be noted that too narrow a weld may be the cause of its poor quality, due to shrinkage cracks or cracking issues involved with weld depth to width ratio (W_D/W_W). This is especially true in high-heat welding processes or higher strength material – or a combination of both. Increasing the cross-sectional area of weld has a direct impact on the cost, and it also has an impact on the quality of the weld. A well-balanced compromise between these two extremes is required. A small groove angle in steel of thickness $\geq$ 16 mm (0.630 inch), and selecting double-groove welds is the best approach. In the sketches in Figure 2-5-3, the impact of the changes in the joint design is highlighted. Note that the change in the groove angle significantly reduces the weld cross-sectional area, requiring less weld metal. The impact of changing design from single-V groove to double-V groove is similar. When designing weldments, a welding engineer must weigh up these options and decide on the best one for the project.

The impact of the size of the weld is particularly significant for fillet welds. It is common practice, mostly due to ignorance and poor understanding of welding, to over-weld or over-size a fillet weld. The effect of over-sizing is illustrated in the following sketch, where we can easily see that a change in welding size dramatically increases the amount of deposited weld metal. The sketch shows that the specified weld size is 6 mm, and the over-deposited weld size is 9 mm; an increase of 50%. This change means that twice as much deposited weld metal was used, compared to the originally sized weld. Such a cost escalation can easily turn a profitable project in to a loss-making project. This emphasizes the need for control of the weld size.

The accessibility of the weld joint is another important factor to consider while designing weldments. Often the practice of providing a 'weld access hole' is followed when a weld is designed in a very tight location. These access holes are sized so as to allow for proper clearance for the deposition of sound weld metal. The welding engineer responsible for designing such weldments should be able to recognize the need for these access holes, and decide on the minimum required size of the hole. Such decisions must take into consideration how the size of hole would affect the maximum net area available in the connected base metal, and must compensate for any loss in effective strength.

Equally important is the proper selection of material for fabrication. A balance between the designed strength requirements and availability must be established. A material, however versatile in theory for a specific project, may be too expensive if not easily available.

And the last but not least important step in cost control is the selection of a suitable welding process, making the best use of available resources.

The above are general guidelines and may be used to assess the actual project conditions. Ignorance of the above, or failing to consider these points for cost-effective fabrication, is not an option for a successful welding engineer.

Control of Welding Stresses to Minimize Through-Thickness Failures

Base metal through-thickness loading, especially by welding stresses, is a very important factor. The weld joins in 'T' or corner joints configurations are especially subject to such loading, and likely to fail due to through-thickness stresses. The function of these type of joints is to transmit stress normal to the surface of a connected part, especially when the base metal thickness of the branch member or the required weld size is equal to or exceeds 20 mm (0.75 inch). Such configurations must receive special attention in design, when base metal selection and details of the weld joint should be considered carefully. Joints which minimize stress intensity on base metal through-thickness direction should be used. Control of the specified weld size as discussed above assumes critical importance in such designs. Control on stress is of utmost importance in avoiding through-thickness failures of any member of a weldment.

Chapter 6

Welding Corrosion Resistant Alloys – Stainless Steel

In previous chapters we discussed the principles of welding, types of equipment, various welding processes and the effects of heating and cooling on metals. We used some examples to illustrate the points discussed; most of these examples

Applied Welding Engineering: Processes, Codes and Standards.

were taken from common experiences, and used a common material – carbon steel.

In this and subsequent chapters we will concentrate on a specific material group, commonly known as corrosion resistant alloys (CRAs). Stainless steel of various grades comprises the largest part of the CRA group. In this chapter we shall focus on the metallurgical and mechanical properties of these materials from the welding engineer's point of view. In the discussion, we shall apply the principles and physics of welding learned so far.

CORROSION RESISTANT ALLOYS (CRAs)

The term corrosion resistant alloy (CRA) is commonly used to cover all metals that are capable to some degree of resisting corrosion, as compared with carbon steel. This resistivity to corrosion is not universal as it depends to a large degree on the specific corrosion environment and on the alloying composition of the given alloy.

The alloys vary considerably from each other and can be iron or nickel-based materials, often alloyed with chromium or copper or any possible variations of different alloying elements. In fact all the metals discussed in this chapter, and some in subsequent chapters, are part of the large and diverse family of CRA materials.

STAINLESS STEEL

Stainless steel is chosen for a project based on any single or a combination of the following specific properties:

1. Resistance to corrosion
2. Resistance to oxidation at higher temperatures
3. Good mechanical properties at room temperature
4. Good mechanical properties at low temperature
5. Good mechanical properties at high temperature
6. Aesthetic values – good appearance.

Stainless steels are corrosion resistant materials which rely on surface passivity for their resistance to corrosion attack. Use of these materials is governed by the oxidizing characteristics of the environment. For more oxidizing conditions, stainless steels are superior to several of the more-noble metals and alloys that are available for fabrication and by welding.

WELDING STAINLESS STEEL

General Welding Characteristics

All the Chromium-Nickel (300 Series) austenitic stainless steels, with the exception of high sulfur or selenium free-machining grades (AISI 303), are easily welded. The welded joints are tough and ductile in the 'as welded' condition. These welds, if used in non-corrosive or mildly corrosive services, do not require any post-weld heat treatment.

In welding, a temperature gradient is achieved ranging from room temperature to molten steel. The area that is heated in this process ranges from 425°C to 900°C (800°F to 1,650°F) and becomes sensitized, as carbides are precipitated. This carbide precipitation may affect the life of equipment under severely corrosive conditions, therefore to restore optimum corrosion resistance annealing the welded parts is recommended. This annealing process is called solution annealing, and consists of heating the material up to a temperature above the sensitizing temperature – generally 1,100°C (about 1,960°F) – and holding it for long enough for the carbon to go into solution. After this, the material is quickly cooled to prevent the carbon falling out of the solution. Solution annealed material is in its most corrosion-resistant and ductile condition.

It is not always possible to solution anneal the weldments. This could be due to its size, or other post-fabrication process etc. If for any reason the welded part cannot be annealed, then extreme care should be taken in welding stainless steel, and either a low carbon grade of stainless steel with less that 0.03% carbon, or AISI 321 or 347 grade should be selected. The latter are stabilized alloys containing titanium and columbium respectively. The ratio of these elements is dictated by the percentage of carbon in these steels. For example the minimum amount of titanium in Grade 321 is about 5 times that of the carbon in the steel. Similarly columbium in Grade 347 is present at about 10 times the carbon content of the steel. The exact ratio of carbon to titanium or columbium has to be designed in the steel, based on the specific requirement of the service environment of the project, including its welding requirements. When these steels are heated during welding, and the material reaches the sensitizing temperature range, carbide precipitation occurs, as in any other grades of stainless steel. The high affinity of carbon to these added elements means that the carbide of titanium or columbium is precipitated, thus leaving the chromium in the solution and free from intergranular corrosion. In some very specialized conditions Grade 321 may be further heat treated by heating to 815°C–900°C for 2 to 4 hours, and air cooled to secure complete carbide precipitation as stable titanium carbides. This heat treatment is sometimes called stress relief treatment.

In low carbon (less than 0.03%) grades like 304L and 316L, the carbon level is so low that the heat does not precipitate carbides during welding. The use of these grades of steel is limited to service temperatures below 425°C–870°C (800°F–1,600°F).

Welds in other corrosion-resistant steels, such as ferritic and martensitic stainless steels, are not as ductile and tough as the austenitic steels discussed above. Ferritic alloy types 405, 430, 442, and 446 are more readily weldable. The martensitic grades, like 403 and 410, are more weldable than types 420 and 440 grades in the same group.

Welding Processes

All arc welding processes (including electron beam, laser beam, resistance, and friction welding) readily join stainless steels. Gas metal arc, gas tungsten arc, flux cored arc, and shielded metal arc welding are also commonly used welding processes. Plasma arc and submerged arc welding (SAW) are also suitable joining methods.

Restrictions on SAW are necessary because chemical composition control of the weld deposit is more difficult due to the effect of arc voltage variations. Heat input is higher, and solidification of the weld metal is slower, which can lead to large grain size and lower toughness. Ferrite contents of at least 4% are nearly inevitable. However there are several commercial developments in process and consumables that address these concerns to some degree which are now emerging in the market. Closer scrutiny of the welding procedure is advised in these applications.

Oxyacetylene welding is not recommended due to the high heat input required.

Protection against Oxidation

A welding process must protect the molten weld metal from the atmosphere during arc transfer and solidification. Fluxing may be required to remove the chromium and other oxides from the surface and the molten weld metal. Gas shielded processes do not require fluxing, since the shielding gas prevents oxidation.

Welding Hygiene

Importance of Cleaning Before and After Welding

The high chromium content of stainless steels promotes the formation of tenacious oxides that must be removed for good welding results. Surface contaminants affect stainless steel welds to a greater extent than they affect carbon steel welds.

The surfaces to be joined must be cleaned prior to welding. A band of at least 12 mm (0.5 inch) surrounding the weld joint is cleaned, and this band must be far bigger if thicker plates are being welded. As a general rule of thumb, cleaning a band of metal about 1.5 times the plate thickness is considered good practice, and will avoid contamination.

Special care in surface cleaning is required for gas shielded welding because of the absence of fluxing.

Carbon contamination can adversely affect the metallurgical characteristics and corrosion resistance of stainless steel. Pickup of carbon from contaminants or embedded particles must be prevented.

Suitable solvents are used to remove hydrocarbon and other contaminants such as cutting fluids, grease, oil, waxes, and primers. Light oxide films can be removed by pickling or by a carefully selected mechanical means of cleaning. Acceptable pre-weld cleaning techniques include:

1. Stainless steel wire brushes that are used only for this purpose.
2. Blasting with clean sand or grit.
3. Machining and grinding with chloride-free cutting fluid.
4. Pickling with 10% to 20% nitric acid solution.

Thorough post-weld cleaning is required to remove welding slag. The surface discoloration is best removed by wire brushing or mechanical polishing.

Filler Metals

Covered electrodes and bare solid and cored wire are available to weld most of the grades. The chemical composition of all-weld metal deposits varies slightly from the corresponding stainless steel metal composition to ensure that the weld metal will have the desired microstructure and be free from cracks.

Covered electrodes are available with either lime or titania coverings. Lime type coverings (Exxx-15) are only suitable for DCEP (electrode positive, reverse polarity) and Exxx-16 electrodes are suitable for AC or DCEP. Type Exxx-15 electrode coverings produce deeper penetration and Type Exxx-16 electrode coverings produce a smoother surface finish when used with a DCEP current polarity system.

Covered electrodes must be stored in sealed containers or holding ovens at temperatures of 100°C to 125°C (200°F to 250°F).

AUSTENITIC STAINLESS STEELS

Metallurgical Concerns Associated with Welding Austenitic Stainless Steels

The properties of austenitic stainless steel which are valuable to industry are; high ductility, excellent toughness, strength, corrosion resistance, weldability, and excellent formability and castability. Because of these properties, austenitic stainless steels are the most commonly used material within the family of stainless steels. There is a virtual continuum of austenitic alloys containing Fe, Cr, Ni, and Mo.

The distinction between highly-alloyed stainless steels and lower-alloyed nickel-based alloys is somewhat arbitrary. Nickel alloys must satisfy either (a) Cr >19; Ni >29.5; Mo >2.5, or (b) Cr >14.5; Ni >52; Mo >12 over their entire composition range. Those that do not meet this criteria, e.g., alloy 20, UNS N08020 are classified as stainless steels.

Unified Numbering System (UNS) alloy numbers starting with a prefix 'S' are grouped with the austenitic stainless steels discussed in this section, whilst superaustenitic stainless steels, defined in this section as alloys with FPREN greater than 30.0 are discussed in the section entitled 'Superaustenitic

Stainless Steels'. The alloys that begin with prefix "N' are grouped with the nickel-based alloys.

Mechanical Properties of Stainless Steels

The lower-alloyed austenitic stainless steels, such as types 304 and 316 (UNS S30400 and S31600), possess yield strengths of around 30 to 40 ksi (210 to 280 MPa) in the annealed condition. Some higher-alloyed austenitic stainless steels with nitrogen have higher yield strengths.

Cold working often increases strength, especially in higher-alloyed austenitic stainless steels.

Cold deformation during fabrication, although less severe than that applied during temper rolling, can produce martensite in some austenitic stainless steels, thereby increasing their susceptibility to hydrogen embrittlement. Fabrication processes can also induce residual stresses that may increase the possibility of stress corrosion cracking (SCC).

Many of the common austenitic stainless steels can be readily welded using matching filler metals. Higher alloy grades are normally weldable, but non-matching, over-alloyed nickel-base filler metals are used.

Generally, these alloys are readily weldable via a range of processes (SAW, GTAW, GMAW, SMAW, PAW etc). They are usually welded with matching composition filler metal. For some of the molybdenum-containing grades, an over-alloyed filler with an extra 1–3% molybdenum and higher nickel content is specified. Normally, argon is used as both shielding and backing gas. Austenitic stainless steels typically require care in welding and adherence to good stainless steel welding practices and weld hygiene.

Since these alloys are in the austenitic phase, and they do not undergo phase transformation on cooling, they do not require pre-heat or post-weld heat treatment (PWHT), except in some specific cases where solution annealing may be specified after welding and hot working.

Welding for the typical austenitic stainless steels is commonly practiced using standard consumables like ER308L, etc. Welding of higher-strength (650–690 MPa UTS), 200-series austenitic stainless steels can be welded with standard 308-L-type fillers-metals if matching the strength of the base metal is not critical. Use of duplex ER2209 filler metal is one way of matching or exceeding the strength of the base metal, but toughness and embrittlement concerns restrict the use of this approach to service temperatures of about −40°C to 315°C (−40°F to 600°F). For cryogenic applications, such as liquefied natural gas (LNG) equipment, use of less standard fillers such as E16-8-2 or 316-L Mn, or use of nickel-base fillers such as UNS N06022, are often selected to give the necessary strength and cryogenic toughness.

Welding of Austenitic Stainless Steels

At the beginning of this chapter we discussed the general requirements of welding, and addressed essentials like sensitization control, differences in welding stainless steel and carbon steel and the importance of weld hygiene. We will take those discussions further in the following paragraphs.

Austenitic steels have high coefficients of thermal expansion and low thermal conductivity, and are particularly susceptible to distortion during welding. They have better ductility and toughness than carbon steels and excellent notch toughness even at cryogenic temperatures. They are stronger than carbon steels above 500°C (1,000°F) and have good oxidation resistance.

When austenitic stainless steels are joined to carbon steel, construction codes often mandate PWHT in the temperature range of about 550–675°C (1,025–1,250°F) for relief of residual stresses. These heat treatments can adversely affect the intergranular corrosion and stress corrosion cracking (SCC) resistance of the stainless steel. In these situations, use of a low-carbon grade type 304-L or stabilized grade like type 347-L is recommended. It may however be noted that the service temperature range of 304-L and 347-L is limited to −40°C to 315°C. If PWHT is one of the limiting factors imposed on the design, then other alternatives must be considered, one of which is to butter the carbon steel as described below.

A buttering layer of austenitic stainless steel electrode/filler wire is deposited onto the carbon steel. Often the selection is based on the available chromium in the as-deposited weld metal, after compensating for dilution. If the resulting weld metal is close to the austenitic level when the buttering is completed, the most common interface electrode for welding and buttering austenitic steel and carbon steel is E 309 grade of consumable. Once the buttering is completed, the buttered carbon steel is heat treated as required. After the post-weld heat treatment (PWHT) is carried out, then the stainless steel member is welded on to the PWHT buttered section of carbon steel.

Heat input ranges and interpass temperatures are not especially important for austenitic stainless steels. Interpass temperatures up to 150°C (300°F) are usually permissible.

After welding, there is usually a heat tint in the weld/HAZ area, and it is usual to remove this where feasible. The heat tint is often removed by manual (but not mechanical) brushing, by mechanical abrasives, such as a flapper wheel, or by a suitable pickling paste or gel. The inside of small bore pipe welds, flowlines, and clad line pipes are difficult to clean; this requires that the welding procedure uses inert gas as backing to keep the internal surfaces oxide free.

Weld deposit microstructures are very different from wrought metals of the same composition. In welds 100% austenitic structure is prone to cracking, and some amount of ferrite is essential to control this. In austenitic welds, small pools of delta ferrite often form and carbides may also be present. The weld metal ferrite control is essential; Schaeffler and DeLong diagrams are used to predict as-welded microstructures. These diagrams are also useful in selecting the electrode for controlling the ferrite in austenitic steel weld metal.

SUPERAUSTENITIC STAINLESS STEELS

Material Properties and Applications

Like the austenitic stainless steels, the superaustenitic stainless steels are highly ductile; they have excellent toughness, high strength, outstanding corrosion resistance, good weldability, and excellent formability. They are normally used where greater resistance to corrosion, especially protection from chloride pitting and crevice corrosion, is needed. Superaustenitic stainless steels are defined as austenitic, iron-based alloys that have PREN greater than 40.

The higher PREN values are achieved primarily by adding nitrogen (N) to these alloys, and upper working temperature limits of 400°C (750°F) are generally imposed by industry codes to prevent Σ (sigma) or X (chi) phase embrittlement.

Many of the superaustenitic stainless steels, especially those containing nitrogen, possess higher yield strengths in the annealed condition than the standard austenitic stainless steels.

These alloys are generally available in most product forms (bar, wrought plate, castings, pipe, forgings, etc.), and are usually supplied in the solution-annealed condition. Specialized parts (fittings, fasteners, etc.) of these grades are not generally inventoried by stockists, but are often custom manufactured. Suitable nickel alloys are often selected as an alternative. Castings are solution-annealed to homogenize the as-cast, cored, dendritic structure.

Superaustenitic stainless steels are generally used in the solution-annealed and rapid-cooled condition. Prolonged heating in a temperature range of about 510 to 1,070°C (950 to 1,960°F) can cause precipitation of carbides, nitrides, or intermetallic phases. This precipitation increases their susceptibility to intergranular corrosion, IGSCC, and chloride pitting and crevice corrosion. These alloys cannot be strengthened by heat treatment.

The higher alloy content of the superaustenitic stainless steels gives them greater resistance to the formation of martensite during cold working. Thus, they show reduced susceptibility to hydrogen embrittlement (HE) and greater resistance to stress corrosion cracking (SCC). Fabrication-induced residual stresses are less likely to cause SCC in these alloys.

Welding and Joining of Superaustenitic Stainless Steels

These alloys are easily weldable by a range of processes like SAW, GTAW, GMAW, SMAW etc. All of these processes have been discussed in this book. Because these alloys rely in part on molybdenum to provide their corrosion resistance properties, the segregation of molybdenum that occurs during solidification of welds can impair the corrosion resistance of welds. To counter this effect, normal practice is to use over-matching composition filler metal. The over-alloyed fillers typically contain about 1.5 times the molybdenum than the base metal. To keep these high levels of molybdenum in solid solution, nickel-based fillers are used. This ensures that even the solute-depleted dendrite cores will have local PREN values which meet or exceed the PREN of the base metal. Examples of such filler metals include UNS N06625, N06022, and N06686 wires.

From this description of the weldability of these molybdenum-alloyed steels, it can be seen that autogenous welding is not normally carried out, although it has been performed successfully in thin sections of <2 mm and with special gases. Some specialized post-weld solution annealing process can also restore autogenous welds to corrosion resistance levels approaching that of the base metal. Normally, argon is used as the shielding gas, but the addition of small amounts of nitrogen is considered more beneficial. Backing gases can be argon or nitrogen. As with high alloy stainless steels, care is typically needed in welding superaustenitic steels, and adherence to good stainless steel welding practice is required. Suitable joint design, interpass temperatures and low heat inputs is the path to successful welding of these alloys. As for stainless steel welding, pre- and post-weld heat treatments are not required for superaustenitic stainless steels.

During welding, heat input ranges and interpass temperatures are very carefully monitored and controlled. The maximum permissible heat input and interpass temperature increase with section thickness. The values for these parameters generally decrease as the alloy content increases. Specialist publications giving suitable values for a specific joint should be consulted. If heat input or interpass temperature is kept too high, the risk of precipitating sigma or chi phases in the HAZ or weld metal increases. These intermetallic phases are rich in chromium and molybdenum, thus leaving a chromium-depleted area around them, which locally reduces corrosion resistance.

Austenitic stainless steels contain a combined total chromium, nickel, and manganese content of 24% or more, with the chrome generally above 16%. Nickel and manganese stabilize austenite to below room temperature.

The ferrite content is designated by ferrite number (FN). Ferrite is difficult to measure accurately, although automated equipment is now available. The importance of ferrite in weld microstructure cannot be understated, as it increases resistance to hot cracking. Ferrite provides sites with good ductility for interstitial or tramp elements to distribute. However, excessive ferrite can also lower corrosion resistance and worsen the high temperature properties of material.

Welding parameters and techniques have a significant effect on the amount of ferrite formed and retained in a weld and they must be controlled to produce the desired properties in the weld.

Difficulties Associated with Welding Stainless Steel

Austenitic stainless steel welding appears to be similar to the normal carbon steel welding. But a little more in-depth observation reveals considerable differences, basically due to the metallurgical differences between the two

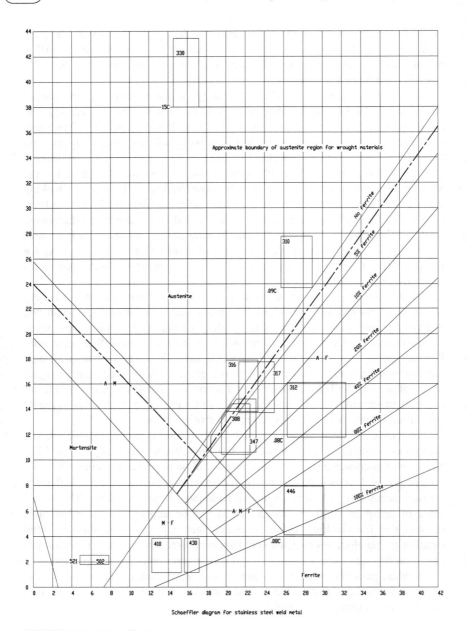

Schaeffler diagram for stainless steel weld metal

FIGURE 2-6-1 Schaeffler diagram.

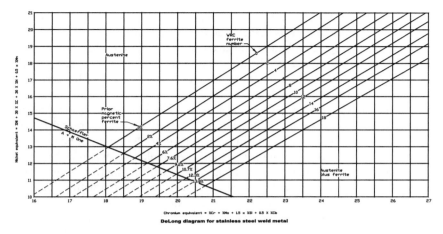

DeLong diagram for stainless steel weld metal

FIGURE 2-6-2   DeLong diagram.

types of steel. Stainless steel does not undergo the normal phase changes associated with carbon steels. While ferritic steels are austenitic and non-magnetic
at elevated temperatures, they transform to ferrite, pearlite, martensite and
other phases as they are cooled through the transformation range.

In contrast, when stainless steel is cooled, all or nearly all of the material
retains the austenite structure at room temperature. No phase changes occur,
and no hardness increase is associated with cooling. This property of austenitic
steels reduces the need to pre- or post-heat.

When these steels are welded, two very specific points need to be
considered:

1. Carbide precipitation or sensitization.
2. Micro-fissuring, ferrite content and sigma phase formation.

The corrosion resistance of austenitic steels depends on the addition of various alloys of which chromium is of primary importance.

In our earlier discussion on stainless steel welding we introduced terms
like sensitization, sigma phase etc. We will now discuss these in a little more
detail. When austenitic steel is heated to a temperature range called the 'sensitization range' – i.e., 425°C to 870°C (800°F to 1,600°F) – some of the chromium in the solution can combine with any carbon that is available and form a
chromium-rich precipitate of chromium carbide, thus reducing the chromium
in the steel. Less chromium is now available in the alloy to carry out its primary duty; that is to resist corrosion. This process is called sensitization.

Steel in such a condition is easily attacked by an acidic environment.
Under certain conditions, austenitic welds are subject to intergranular corrosion. A narrow band of metal in the heat-affected zone is always heated to the

sensitizing range. The amount of precipitation that occurs is a function of the carbon content. Carbon levels higher than 0.03% are particularly susceptible.

Many metallurgical solutions are available to overcome the formation of sensitized area in austenitic welds; we will discuss them in the next paragraphs.

Since we have discussed the importance of austenite in steel for corrosion resistance, we should make it clear that, while pure austenite has excellent mechanical and corrosion resistance properties, its ability to absorb impurities without cracking during solidification is severely limited. During the cooling process the low-melting impurities are forced out to the grain boundaries. Excessive amounts of such grain boundary accumulations weaken the material at that point, creating grain boundary flaws called 'micro-fissures'. This is significant because of the very small area that is associated with welding in relation to the parent metal. One method to reduce such micro-fissuring is to disperse these impurities among the disconnected grain boundaries that surround the island of second phase. This can be accomplished by modifying the chemical composition of steel in such a way to allow the creation of islands of ferrite in the welds. Ferrite has an enormous capacity to absorb impurities, and ferrite islands are dispersed throughout the microstructure. The presence of ferrite has the potential to reduce the corrosion resistance of steel very slightly, however it can certainly prevent micro-fissuring, which is more serious and can lead to catastrophic failures.

On the other hand, too much ferrite is also detrimental to the material. It can cause other problems due to a 'sigma phase' developed within the welding temperature range. This is a very brittle constituent and is caused by very evenly dispersed ferrite. Even a small amount of sigma phase will embrittle large areas of stainless steel.

It is clear that both minimum and maximum limitations on ferrite phase are desirable in stainless steel welding to prevent micro-fissuring and sigma phase embrittlement.

Since carbon rapidly decreases the corrosion resistance and changes the properties of austenitic welds, it must be carefully controlled. Filler metals are usually chosen to match the base metal composition.

We have established that the ferrite content must be appropriate for the weldment's service requirements. Many electrodes have been developed to produce deposits containing ferrite limits within the range of 4 to 10 'ferritic number'. In addition, there are methods that can measure ferrite in the weld; one of them is magnetic ferrite gauge as described in AWS A 4.2.

MARTENSITIC STAINLESS STEELS

Properties and Application

Martensitic stainless steels are Fe-Cr-C alloys that are capable of the austenite-martensite transformation under all cooling conditions. Compositions for most martensitic steel alloys are covered by a number of specifications, such as ASTM A 420 or API 13 Cr L80 and 420 M with additional small amounts of Ni and/or Mo. Although 9Cr-1Mo is not strictly a martensitic stainless steel, it is often included in this alloy group, because the challenges associated with the welding of 9-Cr-Mo steel is in many ways similar to this group of materials.

The martensitic stainless steels are generally used in the quenched and tempered, or normalized and tempered condition. For services where hydrogen evolution or the presence of sulfur is expected, as in sour gas services in the oil and gas industry, a maximum hardness of 22 HRC is usually specified, and for most of the alloys. Some of the alloys like type 410 and 420 develop quench crack if quenched in water, so they are quenched only in oil or polymer, or they are air-cooled before tempering.

Some alloys, like type 410, type 415, and J91540 (CA6NM) receive a second temper treatment called 'double tempering' at a temperature lower than the first tempering temperature, to reduce the untempered martensite in type 410, type 415, and J91540 (CA6NM). Double tempering has not been shown to improve resistance to stress corrosion cracking in type 420 tubular products or for 9Cr-1Mo tubular or forgings.

The mechanical properties of typical base metal strength (SMYS) are grouped as 414 MPa (60 ksi), 517 MPa (75 ksi), 552 MPa (80 ksi), and 586 MPa (85 ksi), with hardness controlled to the maximum 22 or 23 HRC. They often have a specified maximum yield strength of 95 to 100 ksi (660 to 690 MPa). For sour service applications, tubular products are generally used according to the API specification 5CT or L-80 strength level; forgings and castings are generally specified with hardness not exceeding 22 on the Rockwell C scale. Higher strengths are used in sweet service applications; however, corrosion resistance and ductility is adversely affected as the strength of steel is increased.

Welding Martensitic Stainless Steels

The martensitic stainless steels are easy to work with, and the welding processes used include SMAW, GMAW (MIG/MAG), FCAW, GTAW (TIG), SAW, EBW, and laser beam welding (LBW). Typical welding consumables include 410 Ni Mo (matching weld metal), or 2209, 309LSi (overmatching consumables), while some limited application of autogenous welding is also practiced. These alloys are not used in the as-welded condition in more demanding environments, like sour service.

Extreme care is typically required when these alloys are welded, because they are susceptible to high hardness. Tubing and casing are generally not welded.

When welding type 410, high pre-heat temperatures are used. The alloys classified as type 410, type 415 (F6NM), and J91540 (CA6NM) are tempered again as a post-weld heat treatment, to ensure that they have maximum specified strength and hardness. These alloys have been welded using nominally matching filler metals. The use of non-matching, austenitic type consumables can increase the risk of fusion boundary cracking in sour service; this is irrespective of the hardness limits in the weld area.

These alloys are known for moderate corrosion resistance, heat resistance up to 535°C (1,000°F), relatively low cost, and the ability to develop a wide range of properties by heat treatment.

If left in the as-welded condition, intergranular (sensitization) cracking is a common occurrence in both sweet (CO_2 containing) and sour conditions. These problems also arise as a result of poor PWHT cycles, where the treatment has been ineffective in refining the structure and reducing the HAZ hardness.

They can be air hardened at temperatures above 815°C (1,500°F) for nearly all section thicknesses. Maximum hardness is achieved by quenching above 950°C (1,750°F). They lack toughness in the as-hardened condition and are usually tempered.

Martensitic alloys can be welded in any heat treating condition. Hardened materials will lose strength in the portion of the heat affected zone. With a carbon content of 0.08% and 12% Cr (Type 410), the heat affected zone will have a fully martensitic structure after welding. The steep thermal gradients and low thermal conductivity combined with volumetric changes during phase transformation can cause cold cracking.

The hardness of the heat affected zone depends primarily on the carbon content and can be controlled to some degree by developing an effective welding procedure. As the hardness of the heat affected zone increases, its susceptibility to cold cracking becomes greater, and its toughness decreases.

Weldability improves when austenitic stainless steel filler is used, because it will have low yield strength and good ductility. This also minimizes the strain imposed on the heat affected zone.

Martensitic steels are subject to hydrogen-induced cracking as are low alloy steels. Covered electrodes used for welding must be low-hydrogen and maintained in dry conditions.

Preheating and good interpass temperature control is the best means to avoid cracking. Preheating is normally done in the 200°C to 315°C (400°F to 600°F) range. Carbon content, joint thickness, filler metal, welding process, and degree of restraint are all factors in determining the pre-heat, heat input, and post-weld heat treatment requirements.

Post-weld heat treatment is performed to temper or anneal the weld metal and heat affected zone with the aim of decreasing hardness and improving toughness, and decreasing the residual stresses associated with welding. When matching filler metal is used, the weldments can be quench-hardened and tempered to produce uniform mechanical properties.

Types 416 and 416Se are free machining grades that must be welded with caution to minimize hydrogen pickup. ER312 austenitic filler metal is recommended for welding type 416 and 416Se alloys, since it can tolerate the sulfur and selenium additions.

Type 431 stainless steel can have high enough carbon to cause heat affected zone cracking if proper preheat, preheat maintenance, and slow cooling procedures are not followed.

FERRITIC STAINLESS STEELS

Properties and Application

Ferritic stainless steels are Fe-Cr-C alloys with ferrite stabilizers such as aluminum (Al), columbium (Cb), molybdenum (Mo), and titanium (Ti), added to inhibit the formation of austenite on heating. They are therefore non-hardenable. In annealed conditions, lower-alloy ferritic stainless steels have mechanical properties somewhat similar to the low-alloy austenitic stainless steels, like type 304. The typical yield strength is in the range of 30 to 50ksi (205 to 345MPa). Alloys with increased chromium, molybdenum, and nickel content have higher strengths.

In the high-chromium containing alloys, such as UNS S44626, a welding procedure is typically developed to minimize interstitial pickup during welding, and to retain material toughness. These alloys are predominantly utilized as thin-walled tubing products.

These alloys generally exhibit a loss of toughness with increasing section thickness, and a maximum thickness has been established for each alloy depending on the toughness requirements. In high-chromium containing alloys, the interstitial contents have been carefully controlled for this purpose.

First generation ferritic alloys (Types 430, 422, and 446) are subject to intergranular corrosion after welding, and exhibit low toughness.

Second generation ferritic alloys (Types 405 and 409) are lower in chromium and carbon and contain powerful ferrite and carbide formers to reduce the amount of carbon in solid solution. Although they are largely ferritic, some martensite can form as a result of welding or heat treating. Ferritic alloys are low cost, have useful corrosion resistance and low toughness properties.

Recent improvements in melting practice have resulted in third generation ferritic alloys with very low carbon and addition of nitrogen, e.g., Types 444 and 26-1. Stabilizing with powerful carbide formers reduces their susceptibility to intergranular cracking after welding, improves toughness, and reduces susceptibility to pitting corrosion in chloride environments and to stress corrosion cracking.

The most important metallurgical characteristic of the ferritic alloy is the presence of enough chromium and other stabilizers to effectively prevent the formation of austenite at elevated temperatures. Most grades do form some small amount of austenite since interstitials are present.

Since austenite does not form, and the ferrite is stable at all temperatures up to melting, these steels cannot be hardened by quenching. The small amounts of austenite which may be present and transform to martensite are easily accommodated by the soft ferrite.

Annealing treatment at 760°C to 815°C (1,400°F to 1,500°F) is required to restore optimum corrosion resistance after welding. Ferritic stainless steels cannot be strengthened appreciably by heat treatment. These steels are generally used in the annealed condition. The cooling rate chosen depends on the particular alloy. The choice is important, because the higher-chromium containing alloys are subject to embrittlement by sigma or alpha prime phase if not properly heat treated.

All ferrites heated above 927°C (1,700°F) are susceptible to severe grain growth. This reduces the material toughness, which can only be restored by cold working and annealing.

Welding Ferritic Steel

Types 430, 434, 442, and 446 are susceptible to cold cracking when welds are made under heavy restraint. A 150°C (300°F) preheat can minimize residual stresses that contribute to cracking. These steel grades are also susceptible to intergranular corrosion.

Filler material selection would include any of the three available options:

(1) Matching compositions
(2) Use of austenitic stainless steels consumables, and
(3) Use of nickel alloy consumables.

Matching fillers are normally used only for Types 409 and 430. Austenitic stainless steels electrode or filler wire matching E309 or E312 grade or nickel alloys are often selected for dissimilar welds.

The need for preheating is determined by the chemical composition, desired mechanical properties, thickness, and conditions of restraint. High temperatures can cause excessive grain growth and cracking can occur in the heat affected zone in some grades. Low 150°C (300°F) and interpass temperatures are usually recommended.

If a post-weld heat treatment is deemed necessary, it is done in the 700°C (1,300°F) to 843°C (1,550°F) range to prevent excessive grain growth. Rapid cooling through the 538°C (1,000°F) to 371°C (700°F) range is necessary to prevent embrittlement.

PRECIPITATION HARDENED STAINLESS STEELS

Properties and Application of Precipitation Hardened Steels

Precipitation hardening stainless steels can give them high strength through simple heat treatments. They have good corrosion and oxidation resistance without the loss of toughness and ductility that is normally associated with high strength materials. Precipitation hardening is promoted by alloying elements such as copper (Cu), titanium (Ti), columbium (Cb), and aluminum

(Al). Submicroscopic precipitates formed during the aging treatment increase hardness and strength.

Martensitic precipitation hardened steels provide a martensitic structure which is then aged for additional strength. Semi-austenitic precipitation hardened steels are re-heated to form martensite and also aged. Austenitic precipitation hardened steels remain austenitic after cooling and strength is obtained by the aging treatment.

As a group, the precipitation hardened steels have a corrosion resistance comparable to the more common austenitic stainless steels. Corrosion resistance is dependent on the heat treatment and the resulting microstructure. Welding can reduce corrosion resistance by overaging and sensitization.

Precipitation hardened steels tend to become embrittled after exposure to temperatures above 300°C (580°F), particularly if heated for long periods of time in the range of 370°C to 427°C (700°F to 800°F). After welding, the best mechanical and corrosion resistance properties can be obtained by solution heat treatment followed by aging. For some applications, just the aging treatment is sufficient.

Martensite precipitation hardened steels are often fabricated in the annealed or overaged condition to minimize restraint cracking. Solution heat treatment and aging is performed after fabrication.

Welding Precipitation Hardened Steels

Semi-austenitic precipitation hardened steels are welded in all conditions. Austenitic conditioning and aging is performed after welding for optimum mechanical properties. Austenitic precipitation hardened steels are difficult to weld because of cracking problems. Matching nickel alloy or austenitic filler materials are used. The selection of suitable filler metal is dependent on the post-weld heat treatment and final property requirements.

The following are the key points that must be kept in mind for selection of material as well as welding of all stainless steels discussed thus far:

- Thermal expansion, thermal conductivity, and electrical resistivity have significant effects on the weldability of stainless steels
- The relatively high coefficient of thermal expansion and low thermal conductivity of austenitic stainless steel require better control of distortion during welding
- Low thermal conductivity for all stainless steels indicates that less heat input is required
- The weldability of the martensitic stainless steels is affected mainly by hardenability that can lead to cold cracking
- Welded joints in ferritic stainless steels have low ductility as a result of grain coarsening related to the absence of an allotropic transformation

- The weldability of the austenitic stainless steels is governed by their susceptibility to hot cracking
- Precipitation-hardened stainless steels have welding difficulties associated with transformation (hardening) reactions
- Stainless steels which contain aluminum or titanium can only be welded with gas-shielded processes
- The joint properties of stainless steel weldments will vary considerably as a result of their dependence on welding process and technique variables
- Suitability for service conditions such as elevated temperature, pressure, creep, impact, and corrosion resistance must be carefully evaluated. The complex metallurgy of stainless steels must be considered.

TABLE 2-6-1 Stainless Steel Welding Electrodes and Heat Treatments

| AISI Type | Recommended Heat Treatment | | Common Recommended Electrode for Welding |
	Pre-Weld	Post Weld	
301, 302	Not required if steel temp is above 15°C	Rapid cooling from temperatures between 1065°C to 1150°C (1950°F to 2100°F), if service condition is moderate to severe corrosive.	308
304	As above	Rapid cooling from temperatures between 1010°C to 1095°C (1850°F to 2000°F), if service condition is severe corrosive.	308
304L	As above	Not required for corrosion resistance.	308L, or 347
309, 310	As above	Not required for corrosion resistance, because steel is usually at higher temperature in service.	309, 310
316	As above	Rapid cooling from temperatures between 1065°C to 1150°C (1950°F to 2100°F), if service condition is severe corrosive.	316
316L	As above	Not required for corrosion resistance.	316L

(Continued)

TABLE 2-6-1 (Continued)

AISI Type	Recommended Heat treatment		Common Recommended Electrode for Welding
	Pre-Weld	Post Weld	
317	As above	Rapid cooling from temperatures between 1065°C to 1150°C (1950°F to 2100°F), if service condition is severe corrosive.	317
317L	As above	Not required for corrosion resistance.	317L
321,347	As above	Not required for corrosion resistance.	347
Ferritic and Martensitic Steels			
403,405	150°F to 300°F Light gauge sheet needs no preheat	Air Cool from 1200°F/1400°F (650°C/760°C).	410
410	As above	Air Cool from 1200°F/1400°F (650°C/760°C).	410
430	As above	Air Cool from 1400°F/1450°F (760°C/785°C).	430 Can be welded with 308, 309 or 310 without preheat.
442	As above	Air Cool from 1450°F/1550°F (785°C/840°C).	446
446	300°F to 500°F	Rapid cooling from temperatures between 840°C to 900°C (1550°F to 1650°F).	446
501	300°F to 500°F	Air Cool from 1325°F/1375°F (715°C/745°C).	502
502	300°F to 500°F Light gauge sheet needs no preheat	Air Cool from 1325°F/1375°F (715°C/745°C).	502 Can be welded with 308, 309 or 310 without preheat.

DUPLEX STAINLESS STEELS

This alloy group has been developed over the past 30 years; the development progress has resulted in a range of compositions including lean 22% chromium (Cr) and 25% chromium, as listed in Table 2-6-2 below. These alloys have high strength, good toughness, good corrosion resistance, good weldability, and formability, all of which ease manufacturing. These alloys combine the strength characteristics of ferritic stainless steels and the corrosion resistance of austenitic stainless steels. They have higher resistance to environmental corrosion than austenitic stainless steels. Dual phase alloying requires lower Ni and Mo contents than single-phase austenitic alloys.

The alloys with higher F_{PREN} values are possible through the addition of nitrogen to the alloy.

Duplex stainless steels contain up to 22% chromium. Their key property of value to industry is the material's pitting resistance F_{PREN}, which is typically in the range of 35 to 40.

The chromium content of super duplex steel is up to 25% and its pitting resistance F_{PREN} is typically in the range of 40 to 45.

Mechanical Properties

The mechanical properties of the different types of duplex stainless steel are shown in Table 2-6-3 below. The mechanical properties of the cast versions of these alloys (e.g., UNS J93380, J92205, etc.) are lower than their wrought counterparts. ASTM A 995 'Standard Specification for Castings, Austenitic-Ferritic (Duplex) Stainless Steel, for Pressure-Containing Parts' details the compositions and mechanical properties of cast duplex alloys that are used for pressure-containing parts.

The duplex stainless steels used by the oil and gas industry have a roughly 50/50 austenite/ferrite composition; in general duplex steel of various types presents a phase balance in the range 35% to 65% ferrite. They have adequate toughness at low temperatures, and the alloy is commonly used at temperatures as low as −60°C (−76°F).

Superduplex stainless steel (UNS S32760) has been successfully used down to −120°C (−184°F), but this requires well-developed welding procedures and closely monitored welding parameters during the production process.

On long exposure to temperatures above 320°C (608°F) and up to about 550°C (1,022°F), the ferrite decomposes to precipitate alpha prime. This phase causes a significant loss of ductility; hence, duplex stainless steels are not normally used above 300°C (572°F).

In oil and gas service applications these alloys have fared very well in both sour and sweet environmental conditions.

TABLE 2-6-2 Nominal Compositions of Duplex Steel

Type	UNS No.	Nominal Composition (wt%)								F_{PREN}
		Fe	Cr	Ni	Mo	N	Cu	W	
Lean	S32101	Bal	21	1.5	0.5	0.16	0.5	-	25
	S32304	Bal	23	4	0.3	0.16	0.3	-	26
	S32003	Bal	20	3	1.7	0.16	-	-	>30
Standard	S31803	Bal	22	5	3	0.16	-	-	35
	S32205	Bal	22	5	3.2	0.16	-	-	35
25 Cr	S32550	Bal	25	6	3	0.2	2	-	37
Superduplex	S32750	Bal	25	7	3.5	0.27	0.2	-	>40
	S32760	Bal	25	7	3.5	0.25	0.7	0.7	>40
	S32520	Bal	25	7	3.5	0.25	1.5	-	>40
	S39274	Bal	25	7	3	0.26	0.5	2	>40

TABLE 2-6-3 Nominal Mechanical Properties of Duplex Stainless Steels

Type	0.2% Proof Stress (MPa)	Tensile Strength (MPa)	Elongation (%)
Lean Duplex	450	620	25
Standard Duplex	450	620	25
25 Cr Duplex	550	760	15
Superduplex	550	750	25

Heat Treatment

Generally these alloys are used in the annealed, or annealed and cold worked condition. Prolonged heating at temperatures between 260 and 925°C (500 and 1,700°F) can cause the precipitation of a number of phases, including sigma, which reduces toughness and can reduce SCC resistance. Any prolonged heating below the minimum solution-heating temperature is normally to be avoided. Low-temperature toughness generally decreases with decreasing cooling rates in annealing.

Welding and Fabrication

Cold-worked alloys are usually not welded, because the mechanical strength of the weld would be lower than that of the base metal. Annealed alloys are easily welded. The weld filler metal is chosen to produce a desired volume fraction of ferrite and austenite. Hence, fabrication using autogenous (without filler) metal can result in welds that have poorer mechanical and corrosion-resistance properties. The welding procedure is typically developed to control and balance the ferrite/austenite phase; this is essential to prevent deleterious phases or intermetallics.

These alloys are readily weldable by SMAW, SAW, GTAW, and GMAW processes, among others. Where the weld is to be heat treated after completion, it is usual practice to weld with matching composition filler metal.

In as-welded applications, it is usual to use an over-alloyed filler metal with an extra 2% to 2.5% nickel (Ni). This helps to obtain an austenite/ferrite phase balance of about 50/50, if the weld is cooled rapidly. The lean duplex grades are welded with the filler metal used for 22% Cr duplex stainless steels. Except for thin sheets of up to 2 mm thickness, autogenous welding is normally not recommended for duplex stainless steels.

Normally, argon gas is used for both shielding and backing gases, and welding does not begin until the oxygen level is dropped below 0.1%. As with high alloy stainless steels, care is to be taken in welding duplex alloys, and adherence to the good stainless steel welding practices discussed earlier in this chapter is advisable. Good joint design, control of interpass temperatures and keeping low heat inputs are other essential variables for good welding. Preheat and post-weld heat treatment are not required for duplex stainless steels.

The maximum permissible heat input and interpass temperature increase with section thickness. The values for these parameters generally decrease as the alloy content increases. If heat inputs or interpass temperatures are too high, there is a risk of precipitating sigma (Σ) or chi (X) phases in the heat affected zone (HAZ) or weld metal. These are intermetallic phases, rich in chromium and molybdenum, that leave a denuded area around them, which reduces the localized corrosion resistance. Sigma and chi phases also reduce impact toughness properties. In many applications, especially oil and gas, the low temperature toughness is compromised in favor of corrosion resistance properties.

After welding, there is usually a heat tint in the weld and heat affected zone (HAZ), and it is normal to remove this by manual brushing, by mechanical abrasives, or by a suitable pickling solution or gel.

While developing welding procedures, it is best to include a corrosion test – for example testing according to ASTM G 48, (http://www.astm.org) as part of the weld qualification procedure.

The corrosion test sets important weld parameters, hence it is essential that the qualified parameters of welding are followed very closely during production welding. Experience tells us that sometimes 'less experienced' welders have difficulty in passing the corrosion test. Although the weld made by these less experienced welders meets mechanical requirements, it may not meet the corrosion tests as specified above. This increases the importance of welders'/operators' qualification test and production weld parameters monitored by inspectors. In a very limited way this problem can be resolved by use of 2% nitrogen gas along with argon as shielding gas. The reasons for the corrosion test failure can be due to the development of a third phase, which results from poor supervision and control over the heat input and interpass temperature.

Duplex stainless steel welds usually have lower impact toughness than their parent metals. The welding process used often affects the level of toughness, with GTAW welds being the toughest and SAW being the weakest. The weld-metal toughness is a function of both the heat input and the type of flux used. Experience suggests that a minimum of 70-Joule Charpy impact toughness in the parent metal ensures adequate toughness in a duplex weld, and this is easily achieved when correctly welded. To improve the low temperatures toughness, especially for very low temperature services, it is worth considering the use of nickel alloy filler metals, as long as other properties are not compromised; for example, the nickel alloy weld must have the same strength as the parent duplex stainless steel. A practical example of the above would be the selection of C-276 (UNS N10276) filler metal to improve the impact toughness of cast super-duplex (UNS J93380) at $-120°C$ ($-184°F$) service.

Some specifications for duplex material that are used in undersea environments with cathodic protection (CP) require maximum austenite spacing. In a weld, this cannot be controlled and the result cannot be changed by any heat treatment. However, duplex welds usually have a fine microstructure, and meeting a maximum austenite spacing of 30 μm is usually not difficult. Although the welding in itself does not necessarily degrade the resistance of duplex stainless steel against HISC, the presence of higher stress and stress raisers like weld toe poses a significant problem when uncoated duplex stainless steels or steels with defective coating are exposed to CP under mechanical stress. Failures have occurred as a result of this effect, and guidance to avoid them can be sought from industrial specifications.

Welding Non-Ferrous Metals and Alloys

Non-ferrous metals and alloys occupy an important position among engineering construction materials. A large number of such materials have been developed to provide some specific property. It is not practically possible to discuss all of them in this book; however we will cover some of the more common examples. Knowledge of these will provide a foundation on which an understanding of other materials can be built, and new challenges can be faced with some degree of confidence.

ALUMINUM AND ITS ALLOYS

Aluminum is an element and a metal like iron, but it differs in many ways from iron and its most important alloy steel. For example, aluminum has only one allotropic form so there are no phase transformations. This property can be exploited to control its microstructure.

The main methods by which aluminum can be strengthened include deformation, solution hardening, or by introducing precipitates into the microstructure. From a welding engineer's viewpoint, it is a very useful material; it is easily weldable and relatively easily formed into useful shapes. However, the heat introduced by welding can severely disrupt deformed or precipitation hardened alloys.

Aluminum forms a tenacious oxide film; in practice it is impossible to stop it oxidizing on surfaces exposed to air. It is a challenge to find a way of removing this oxide film, and prevent its formation and reformation during welding. However it can be welded by dispersing the adhering oxide by the action of chemicals, or a welding arc, although fragments of oxide are often entrapped in the weld unless a special welding environment is created. For example, resistance spot welding of aluminum is difficult (though not impossible) because the oxide film can cause uncontrolled variations in surface resistance.

There have been some quite remarkable developments in the joining of aluminum and its alloys. A few of these achievements will be reviewed here, including friction-stir welding, flux-free brazing, transient liquid-phase bonding using a temperature gradient, and the joining of aluminum forms.

THE CONFUSING THING ABOUT ALUMINUM

There are a few characteristics of aluminum metal that must be considered if this material is to be welded with consistent ease and quality. Pure aluminum metal has a melting point less than 650°C (1,200°F), and unlike steel aluminum does not exhibit color changes before melting. For this reason, aluminum does not 'tell you' when it is hot and ready to melt. The oxide or 'skin' that forms so rapidly on its surface has a melting point almost three times that of the pure metal; over 1,760°C (3,200°F). To add to this confusion, aluminum boils at about 1,582°C (2,880°F) – i.e., at a temperature lower than the melting point of the oxide adhering to its surface. The oxide is also heavier than aluminum, so when the oxides are melted, they tend to sink in the molten metal and become trapped. Due to these complex reasons, it is recommended that the oxide film is removed from the surface before starting to weld.

WELD HYGIENE

The practice of good weld hygiene is important for any metal, but for aluminum it can never be over-emphasized.

To weld aluminum, operators must take care to clean the base material and remove any dirt, aluminum oxide and hydrocarbon contamination from oils or cutting solvents left over from previous processes like machining etc. As stated above, there is a significant difference between the melting temperature of the oxide films on the aluminum surface and the melting temperature of aluminum itself. This emphasizes the need for removing the oxide prior to starting welding.

Stainless steel bristle wire brushes or solvents and etching solutions are commonly used to remove aluminum oxides. There is a special way of brushing to prevent entrapment of particles of removed oxides in the 'cleaned' surface. The brush movement must be in one direction, and no crisscross or reverse brushing should be done. Care must be taken to brush lightly; brushing too roughly is likely to imbed the oxide particles further into the work-piece. Cross contamination with other materials should be avoided.

If chemical etching solutions are used, it is important to make sure that the remnants of the chemical are fully removed from the work before application of welding heat.

To minimize the risk of hydrocarbons from oils or cutting solvents entering the weld, they should be removed with a degreaser. Ensure that the degreaser used does not contain any hydrocarbons that may contaminate the material and weld area.

PREHEATING

Preheating the aluminum work-piece can help to avoid weld cracking, but the preheating temperature should not exceed 110°C (230°F). To prevent over-heating, close control of the temperature is advised. The use of a laser temperature indicator is recommended.

In addition, tack welding the ends of the plate to be welded helps keep the heat within the work-piece and makes the preheating more effective.

The Conductivity of Heat

Aluminum is an excellent conductor of heat, and requires a large heat input to start welding; this is because most of the heat is often lost through conduction to the surrounding base metal. After welding has progressed for a while, much of this heat has moved ahead of the arc and pre-heated the base metal to a temperature that requires less welding current than the original cold plate. If the weld is continued farther towards the end of the two plates, there is nowhere for this preheat to go, so it is accumulated in the material to such a degree that any further heat would start to melt the metal. Reducing the welding current is one way to address this condition. However, the welding engineer should recognize these challenges and ensure that such heat-related issues are addressed in their welding and fabrication procedures.

Some aluminum alloys containing silicon exhibit 'hot short' tendencies and they are crack sensitive, but these alloys are successfully welded using a filler metal that contains up to 13% silicon in the alloy. Hot shortness is a property that is manifest over the range of temperatures where the solidifying weld metal together with metal that has just solidified has low ductility, and lacks sufficient resistance to shrinking stresses. Proper choice of filler metal and welding procedures along with smaller weld bead deposits help eliminate such issues. Back step-welding techniques are often used effectively for welding these alloys.

Welding Filler Metals

The metal produced in the weld pool is a combination of filler and parent metals. The deposited weld metal must have the strength, ductility, freedom from cracking, and the corrosion resistance required for the specific design application. Table 2-7-1 lists some of the recommended filler metals commonly used for welding various aluminum alloys.

The maximum rate of deposition is obtained with filler wire or rod of the largest practical diameter, while welding at the highest practical welding current. The wire diameter best suited for a specific application depends upon the current that can be used to make the weld. In turn, the current is governed by the available power supply, joint design, alloy type and thickness, and the welding position.

WELDING ALUMINUM WITH THE SHIELD METAL ARC WELDING (SMAW) PROCESS

Aluminum is readily welded using the SMAW process, although developments made in other processes have reduced the general preference for this process. The recommended minimum thickness for welding aluminum by SMAW is 3.2 mm.

The SMAW welding process uses direct current, with the electrode held in positive polarity (DCEP). The control of moisture in the electrode covering and general cleaning of the welding surfaces are both very important to accomplish good welds. The welding electrodes are removed from hermitically sealed containers prior to welding. Electrodes from previously opened containers are 'conditioned' at 175°C to 200°C (350°F to 400°F) in an oven for about an hour prior to welding. For correct conditioning temperatures and times it is best to follow the electrode manufacturer's guidance.

The material to be welded must be pre-heated to about 110°C to 190°C (230°F to 375°F); this helps to obtain good fusion and also improves weld

TABLE 2-7-1 Typical Aluminum Welding Electrodes

AWS Class	UNS	Si	Fe	Cu	Mn	Mg	Zn	Ti	Be	Al Minimum reminder	Other Total
E1100	A91100	*	*	0.05–0.20	0.05	–	0.10	–	0.0008	99.00	0.15
E3003	A93003	0.6	0.7	0.05–0.20	1.0–1.5	–	0.10	–	0.0008	99.00	0.15
E4043	A94043	4.5–6.0	0.8	0.3	0.05	0.05	0.10	0.20	0.0008	99.00	0.15

Single numbers are maximum values.
*Silicon + Fe should not exceed 0.95%

quality. Pre-heating also helps to avoid porosity in the weld, as it removes any moisture remaining on the material surface. The control of pre-heat temperature is very important, as some grades of aluminum, like 6xxx, lose their mechanical properties significantly if heated above 175°C (350°F). Support of the weld area is often provided, to reduce the risk of collapse due to loss of strength of aluminum when heated.

Post-weld cleaning of the flux is important, as the left over electrode covering can cause serious corrosion damage to the material, which is a situation that is easier to address before it becomes a problem. Cleaning is easier when the part is not in service and easy to handle.

WELDING ALUMINUM WITH THE GAS TUNGSTEN ARC WELDING (GTAW) PROCESS

This welding process is well suited to welding aluminum; the slower welding process is most suited for all position welding.

Aluminum welding is mostly associated with the GTAW process, particularly for metals of a lower thickness. Many other processes, of course, can join aluminum, but for the lighter gauges, GTAW is the most suitable process. The application of this process in the aeronautics industry has long been recognized. The popularity of aluminum in automotive applications has brought GTAW welding to greater prominence. Mechanically strong and visually appealing, GTAW welding is the number one process chosen by professional welders for professional racing teams, and by the avid auto-enthusiast or hobbyist.

TYPE OF CURRENT AND ELECTRODE

Both DC electrode positive and AC currents are used for welding aluminum. Argon gas is the gas of choice for manual welding.

Pure tungsten electrodes, classified as EWP by AWS 5.12, or tungsten zirconium electrodes, classified as EWZr-1 (Tip color – Brown), are the preferred electrodes for AC welding. The advantage of AC welding is that the cycle reversal allows the cathodic cleaning of the surface scales and oxides, which results in good quality welds. The electrode performs best when the tip is hemispherical. This shape is produced by 'balling' – striking an arc outside the work and allowing the tip to heat and ball to the required shape.

Direct current with electrode negative (DCEN) is also used for welding aluminum; this provides deep penetration and a narrow weld bead. This process is good for higher thicknesses. The process does not have the advantage of cathodic cleaning so pre-weld cleaning assumes a greater importance. If helium gas is used this will result in deeper penetration than with argon gas. Helium gas welding will also require that a thoriated-tungsten electrode, classified as EWTh-1 (Tip color – Yellow) or EWTh-2 (Tip color – Red), is used.

Grinding the Tip of the Electrodes

To obtain optimum arc stability, grinding the electrode tip is of utmost importance. The grinding should be done with the axis of the electrode held perpendicular to the axis of the grinding wheel. A dedicated grinding wheel should be used to grind these electrodes, to reduce the chances of electrode contamination from other material particles left on the grinding wheel.

Thoriated electrodes do not ball as readily as pure or zirconiated electrodes, and they maintain their ground tip shape much better. If this electrode is used on AC they tend to split affecting the weld quality.

WELDING ALUMINUM WITH THE GAS METAL ARC WELDING (GMAW) PROCESS

Power Source

When selecting a power source for the GMAW of aluminum, the first consideration should be the method of arc transfer. A choice is to be made between a spray arc or pulse arc metal transfer system.

Spray Arc Transfer

Constant-current (cc) or constant-voltage (cv) machines can be used for spray arc welding. Spray arc takes a tiny stream of molten metal and sprays it across the arc from the electrode wire to the base material. For thick aluminum, which requires a welding current in excess of 350 A, constant-current (cc) process produces optimum results.

Pulse Arc Transfer

Pulse are transfer is usually performed with an inverter power supply. Newer power supplies contain built-in pulsing procedures based on filler-wire type and diameter. During a pulsed GMAW process, a droplet of filler metal transfers from the electrode to the work-piece with each pulse cycle of current. This process produces positive droplet transfer and results in less spatter and faster welding speeds than the spray transfer welding process. Using the pulsed GMAW process on aluminum also gives better controlled heat input, easier manipulation of the weld metal in out-of-position welds, and it allows the operator to weld on thin-gauge material at low wire-feed speeds and currents.

Wire Feeder

The preferred method for feeding soft aluminum wire is the push-pull method, which employs an enclosed wire-feed cabinet to protect the wire from the environment. A constant-torque variable-speed motor in the wire-feed cabinet helps push and guide the wire through the gun at a constant force and speed.

A high-torque motor in the welding gun pulls the wire through and keeps wire-feed speed and arc length consistent.

In some shops, welders use the same wire feeders to deliver steel and aluminum wire. In this case, the use of plastic or Teflon liners help to ensure smooth and consistent aluminum-wire feeding. For guide tubes, the use of chisel-type outgoing and plastic incoming tubes to support the wire as close to the drive rolls as possible prevents the wire from tangling. When welding, keeping the gun cable as straight as possible minimizes wire-feed resistance. Correct alignment between drive rolls and guide tubes is important to prevent the aluminum wire shaving.

Drive rolls specially designed for aluminum must be used. The tension on drive-roll allows for delivery at an even rate of wire-feed into the weld. Control of the tension is important, as any excessive tension will deform the wire, and cause rough and erratic feeding. On the other hand, slack tension results in uneven feeding. Both conditions can lead to an unstable arc and weld porosity.

Welding Guns

Welding guns must be kept clean and free from dirt and spatters. A separate gun liner for welding aluminum is recommended. Both ends of the liner must be held in restraint to eliminate gaps between the liner and the gas diffuser on the gun. This would also prevent wire from chafing.

The lines must be changed frequently to reduce the possibility of collecting abrasive aluminum oxide, which can cause wire-feeding problems.

The contact tip diameter should be approximately 0.4 mm (0.015 inch) larger than the filler wire diameters, to allow for the free movement of wire when the tip is expanded due to heat and reducing the tip diameters which often turns into an oval-shaped orifice. To keep control over heat related expansion and other difficulties, welding guns are water cooled if the welding current exceeds 200 amperes.

WELDING TECHNIQUE

The Push Technique

In aluminum welding, the practice of pushing the gun away from the weld puddle rather than pulling it will result in better cleaning action, reduced weld contamination, and improved shielding-gas coverage. This also controls heat in the parent metal and prevents overheating and collapse at the end of the weld.

Travel Speed

Aluminum welding needs to be performed 'hot and fast'. Unlike steel, the high thermal conductivity of aluminum dictates use of higher amperage (higher heat) and voltage settings and higher weld-travel speeds. If travel speed is too slow, the welder risks excessive burn-through, particularly on thin-gauge aluminum sheet.

SHIELDING GAS

Argon gas, due to its good cleaning action and penetration profile, is the most commonly used shielding gas when welding aluminum. A shielding-gas mixture combining argon with up to 75% helium will minimize the formation of magnesium oxide when welding 5xxx-series aluminum alloys.

WELDING WIRE

Welding filler-wire selection should be a carefully made decision. The melting temperature of the wire and parent metal should be matched. It is common to use welding wire of lager diameter because of heavier feed rate; often wires of 2.5 mm to 1.6 mm (3/64 or 1/16 inch) are used, but the size of wire to be used must be considered depending on the overall weld joint design. Thin-gage material is often welded with 0.9 mm (0.035 inch) diameter wire combined with a pulsed-welding mode, where the feed rate is between 2.5 to 7.5 meters per minute.

Convex-shaped welds: In aluminum welding, crater cracking causes most failures. Cracking results from the high rate of thermal expansion of aluminum and the considerable contractions that occur as the welds cool. The risk of cracking is greatest with concave craters, since the surface of the crater contracts and tears as it cools. Welders should therefore build up craters to form a convex or mound shape. As the weld cools, the convex shape of the crater will compensate for contraction forces.

FRICTION STIR WELDING (FSW)

As was stated above, aluminum typically forms a tenacious oxide film. This oxide film makes it difficult, but not impossible, to weld. For this reason, electric resistance welding of aluminum is difficult though not impossible. As we have seen so far, there are numerous ways, including use of flux, use of inert gas and also the action of electrical polarity to remove this impediment from the welding cycle.

The FSW process involves joining the metal without fusion or addition of filler metal. The process results in strong and ductile welds, and is suitable for welding flat and long components. Frictional heating and mechanical deformation create the weld. The circular rotating tool rotates and the friction produced causes heat of approximately 80% of the melting point of the metal. The heat is primarily generated by friction between the rotating tools, the shoulder of which rubs against the work-piece. Adiabatic heating due to the deformation near the pin is distributed volumetrically. Adiabatic heating is caused by pressure change; there is no heat transfer from the environment. Heat is supplied by friction which, in this process, is used to heat the faying surfaces of the material to be welded. Welding parameters are adjusted so that the ratio of frictional to volumetric heating decreases as the work-piece thickness increases. This allows for sufficient heat input per unit length.

The central nugget region containing the onion ring flow is the most deformed region of the weld. The thermo-mechanically affected zone lies between the HAZ and the nugget, where the grains of the original microstructure are retained, but in a deformed state.

The ability to weld efficiently depends on the design of the tool. More development on this is being done; TWI in Abington, Cambridge, UK is leading in this effort and more details can be obtained from their publications on the subject.

NICKEL ALLOYS

A brief introduction to commercial nickel and its alloys is given here. Generally specific alloy types or grades have proprietary names and compositions – the following is a general description. There is a virtual continuum of austenitic alloys containing Fe, Cr, Ni, and Mo; the distinction between highly-alloyed stainless steels and lower-alloyed nickel-based alloys is somewhat arbitrary.

Based on the UNS numbering system, the alloy numbers starting with a prefix 'N' are grouped with the nickel-based alloys discussed in this section. Those beginning with 'S' are grouped with the austenitic stainless steels. There are several common high-nickel cast alloys used in the oil industry.

Commercially pure (99.6%) and wrought nickel has UNS number N02200, and ASTM B 160-63, 725 and 730. Manufacturers should be consulted for more specific details.

The next group are alloys of nickel and copper, with UNS numbers N 04400, N 04404, N 04405, N 05500 etc. These alloys have excellent corrosion resistance to a broader environment, although some of them are useful in specific environments; e.g., UNS C 71500 has excellent resistance to corrosion in marine environments. Technically these metals are copper alloyed with nickel. These alloys are often called cupro-nickel alloys, since the percentage of copper exceeds that of nickel.

The other groups of nickel alloys with additions of iron (Fe) are particularly designed to work in high temperature services and resist chloride stress corrosion. The UNS numbering is generally N 06600, which is a solid solution alloy of Ni, Cr, Mo, Fe and Nb.

Heat Treatment

Solid-solution nickel-based alloys are generally used in the annealed, or annealed and cold-worked condition. These alloys are not designed for being strengthened by a heat treatment.

Mechanical Properties

The maximum yield strength is governed by alloy composition, the cold-working characteristics of the alloy, the maximum yield strength permitted by the application, and the ductility specified. Room temperature yield strengths can range from about 210 to 1,380 MPa (30 to 200 ksi), depending on composition and the degree of cold working. The minimum yield strength for tubing is generally in the range of 760 to 970 MPa (110 to 140 ksi). Casing and liners often have higher yield strength.

Fabrication

Annealed alloys can be welded using GTAW, SMAW, GMAW, SAW, and FCAW. Cold-worked alloys are usually not welded because the mechanical strength of the weldments would be lower than that of the cold-worked region. The mechanical properties of cold-worked tubing, especially in thicker sections, can vary through the section.

High nickel alloys are more prone to casting defects, such as hot tears, cracking, porosity, and gassing. These defects can appear at any stage of the manufacturing process, like shakeout, heat treating, machining, or final pressure testing. Although wrought high-nickel alloys are routinely welded and some even hard-faced, welding the cast alloys is considerably more difficult. Stringent specifications, developed in close cooperation with the foundry, have typically been used to optimize weldability and casting integrity. For production of a good quality casting, the foundry processes, raw material quality, filler material composition, weld repair procedures, and heat treatment are all closely controlled and monitored.

Precipitation Hardenable Nickel-Based Alloys

Precipitation hardenable nickel-chromium alloys often contain a fair amount of iron (Fe). These alloys are used for their corrosion resistance, higher strength and excellent weldability.

Heat Treatment of Precipitation Hardenable Nickel Alloys

These alloys are usually used in solution-annealed, or solution-annealed and aged, or hot-worked and aged, cold-worked and aged conditions.

In sour gas applications, the heat treatment of UNS N07718 is typically selected to give good toughness, yield strength, and corrosion resistance. A solution anneal followed by a single step is often is used. For sour gas applications UNS N07716 and N07725 are solution-annealed and aged.

Mechanical Properties

At room temperature the specified minimum yield strengths of these alloys varies from 340 to 1,030 MPa (50 to 150 ksi).

Welding

Overall, as with all CRAs, welding hygiene is of paramount importance, particularly with respect to the following issues:

- Cleanliness of the parts to be welded
- Maintenance and monitoring of Interpass temperature
- Accuracy and cleanliness of Bevel profile
- Proper weld bead profile
- Use of appropriate welding consumables
- Control of heat input
- Welder qualification
- Correct stress relief
- Correct shielding gas composition and flow rate
- Correct backing gas composition and flow rate
- When welding dissimilar material welding to consider the resultant microstructure and composition of the final weld joint
- Welding processes like TIG (GTAW), MIG/Pulsed MIG (GMAW), and autogenous welding are typically used.

UNS N07718, N07716, and N07725 can be welded by the GTAW process. Repair welding UNS N07718 in the aged condition is possible, but the possibility of micro-cracking is increased. The recommended post-welding heat treatment is solution annealing and aging.

UNS N07750 is not normally welded or repaired in the aged condition.

TITANIUM ALLOYS

Titanium alloys have relatively high tensile strengths and toughness while being lightweight, corrosion resistant, and have ability to withstand extreme temperatures. The higher cost of the material and the difficulty of working them are the limiting factors in their use.

Titanium and its alloys provide corrosion resistance and a range of mechanical properties. They posses yield strengths in the range of 40 ksi to 170 ksi (276 MPa to 1 172 MPa). The alloys are classified by their microstructure and are identified as (α) alpha, (α-β) alpha-beta, or (β) beta alloys. Each of these alloy systems get their specific properties through the different heat treatment cycles they receive, such as annealing or solution treating and aging.

Alpha (α) alloys have a single-phase microstructure; hence their properties are only influenced by their chemistry.

As the name suggests, the (α-β) alpha-beta alloys are a two-phased microstructure system. Their properties can be influenced by a variety of heat treatment cycles.

Beta (β) alloys are precipitation hardened systems, and they offer a wide range of mechanical properties.

Titanium alloys have extensive use in the aircraft and space industry, and limited use in the oil and gas industry, although their use is expanding as a result of their low density and excellent corrosion resistance in H_2S and chloride environments. Stress joints, packers, safety valves, and other completion components have been manufactured from a variety of titanium grades.

Heat Treatment

As stated above, titanium alloys generally fall into one of three broad categories, based on their crystallographic structures: alpha (α), alpha/beta (α-β), and beta (β). Commercially pure titanium is classified as alpha.

Alpha (α) Titanium

These alloys are generally used in the annealed or stress-relieved condition. They are considered fully annealed after heating to 675 to 788°C (1,250 to 1,450°F) for 1 to 2 hours. Stress relieving can be performed at 480 to 590°C (900 to 1,100°F) for ½ to 1 hour.

Alpha (α) alloys are generally fabricated in the annealed condition. All fabrication techniques used for austenitic stainless steels are generally applicable. Weldability is considered good if proper shielding is employed. Contamination of the weld by oxygen or nitrogen is to be avoided.

Alpha/beta (α-β) Titanium

These alloys are generally used in the mill-annealed or solution-treated plus aged condition. Annealing is generally performed at 705 to 845°C (1,300 to 1,553°F) for ½ to 4 hours. Solution treating is generally performed at 900 to 980°C (1,650 to 1,800°F) followed by an oil or water quench. Aging is generally performed at 480 to 705°C (900 to 1,300°F) for 2 to 24 hours. The precise temperature and time are chosen to achieve the desired mechanical properties. To achieve higher fracture toughness or crack growth resistance, these alloys can be processed to an acicular alpha via working and/or final annealing above the alloy's beta transition temperature.

The weldability of this alloy depends on its composition. Generally, fabrication is carried out at elevated temperatures, which is followed by heat treatment. These alloys have very limited ability to be cold formed.

Beta (β) Titanium

Beta (β) alloys are generally used in the solution-treated and aged condition. Cold working and direct-age treatments can increase the yield-strength of these alloys to above 1,200 MPa (180 ksi). Annealing at 730°C to 980°C (1,350°F to 1,800°F) and solution treating are generally performed to control ductility. Aging is typically carried out at 480°C to 590°C (900°F to 1,100°F) for 2 to 48 hours to obtain the desired mechanical properties. A process called duplex aging is also used to improve the response to aging. The duplex aging is performed in two cycles; in the first the material is heated between 315°C to 455°C (600°F to 850°F) for 2 to 8 hours, followed by the second aging cycle at 480°C to 590°C (900°F to 1,100°F) for 8 to 16 hours.

Beta (β) alloys can be fabricated in similar ways to the alpha (α) alloys discussed above; they can be cold formed in the solution-treated condition. Because of the higher yield strength of beta (β) alloys, increased forming pressure is required. The weldability of these beta (β) alloys is good. Aging is carried out after welding to increase the strength. The welding process produces an annealed condition exhibiting strengths at the low end of the beta alloy range.

Weld Defects and Inspection

WELD QUALITY

Quality is a relative term, so different weldments and individual welds may have different quality levels depending on their service requirements. For welding, quality includes factors such as hardness, chemical composition, and mechanical properties.

Acceptable quality means that a weldment is:

- Adequately designed to meet the intended service for the required life
- Fabricated with specified materials and in accordance with design concepts
- Operated and maintained properly.

Acceptance Standards

The quality of weld and thereby the weldment is designed in the initial process of the engineering design. The design specification or code defines the maximum acceptable level of flaws in the weld, or in other words, what is the maximum length, area, and depth of a certain type of flaw that can be accepted without compromising the functional ability of the weldment. This is possible only when the test methods, their limitations and the ultimate demands of the weldment are known to the designers. Several inspection and testing guidelines and some mandatory specifications are available for reference. They all address the issue within the scope of their mandated technical jurisdictions. For example, ASME B 31.3 for pressure piping, API 1104 for pipeline welding, AWS D 1.1 for structural design and welding are some commonly used specifications. There are also other specifications that deal with specific challenges of the test methods, e.g., ASTM E 317 for evaluating the performance characteristics of an ultrasonic pulse-echo testing system without the use of electronic measuring instruments, European Norm (EN) 10160 is for the ultrasonic testing of steel flat products of thickness equal to or greater than 6mm, using a reflection method. ISO 12094 is for ultrasonic testing (UT) for the detection of laminar imperfections in strip/plates used in the manufacture of welded tubes. ASTM A 578 is a standard for straight-beam ultrasonic examination for special applications, ASME SA 275 method for magnetic particle examination of steel forgings, ASME section specifies most of the non-destructive testing associated with welds and materials used for pressure vessels and components.

Fabrication codes and standards give the minimum requirements to ensure that a welded fabrication will operate safely in service. Standards should not be encroached upon or the acceptance level diluted without sound engineering judgment and backing. Where the work is required to meet a specific standard, the requirements of the relevant specification must be strictly followed, unless written clarification or deviation is obtained from the client or the specification issuing body.

Weld quality is verified by non-destructive examination. Acceptance standards are related to the method of examination. All deviations should be evaluated and the acceptance or rejection of a weld should be based on the acceptance standards. Determination of the overall quality requirements is a major consideration involving design, fabrication, operation, and maintenance.

Discontinuities in Fusion Welded Joints

Discontinuities may be related to the welding procedure, the process, design, or metallurgical behavior. Process, procedure, and design discontinuities affect the stresses in the weld or heat affected zone (HAZ). Metallurgical discontinuities may also affect the local stress distribution and may also alter mechanical or chemical (corrosion resistance) properties of the weld or HAZ.

Discontinuities may amplify stresses by reducing the cross-sectional area. A more serious effect is stress concentration – stresses are concentrated at notches, sharp corners, abrupt changes in profile, and (especially) cracks. Discontinuities should be considered in terms of:

1. Size.
2. Acuity or sharpness.

3. Orientation with respect to the principal working stress.
4. Location with respect to the weld, joint surfaces, and critical sections of the structure.

Because of the serious implications of the above, imperfection sizing is a critical activity that a welding engineer has to know and master, in order to be able to decide on whether a given size and type of discontinuity will cause the failure of the structure. The following is the method used to determine the sizes of discontinuities in welds.

Sizing of Discontinuities

Sizing of discontinuities to determine whether they are acceptable or not is essential. It is one of the responsibilities of the welding engineer. The need for effective sizing and critical measurement of planar discontinuities cannot be over-emphasized. This activity becomes very important when dealing with welds that are subject to stresses causing fatigue, and carbon steel and alloy steel weldments that are in low temperature service, which can lead to brittle fracture. The maximum size of discontinuity to prevent brittle fracture can be determined. To calculate the effective imperfection of size the following parameters can be used:

$$\overline{a} = C\,(\delta/\varepsilon_y) \tag{1}$$

Where:
$\overline{a}$ is the effective imperfection size parameter.

$$C = (1)/2\pi\{\sigma_a/\sigma_y\}^2 \text{ for } \sigma_a/\sigma_y \leq 0.5 \tag{2}$$

or:

$$C = (1)/2\pi\{(\varepsilon_a/\varepsilon_y) - 0.25\} \text{ for } \sigma_a/\sigma_y > 0.5 \tag{3}$$

Where:
σ_a = maximum effective applied tensile bending stress, MPa
σ_y = Specified minimum yield strength of pipe, MPa
ε_a = Maximum effective applied tensile bending strain
δ = CTOD fracture toughness value, mm
ε_y = Elastic yield strain = σ_y/E (E = Young's modulus)

CLASSIFICATION OF WELD JOINT DISCONTINUITIES

Weld discontinuities can be classified based on the metallurgical or design related welding process or welding procedure.

Included in the design group of welding process or procedure related discontinuities are flaws such as:

- Undercut
- Concavity or convexity

- Misalignment
- Improper reinforcement
- Excessive reinforcement
- Burn-through
- Overlap
- Incomplete penetration
- Lack of fusion
- Shrinkage
- Surface irregularity
- Arc strike
- Inclusions like slag or tungsten
- Oxide film
- Arc craters
- Spatters.

In the metallurgical group the following types of discontinuities are found:

- Cracks or fissures, which may be hot cracks, cold or delayed cracks, stress cracks or cracks due to reheat
- Lamellar tearing
- Porosity: spherical, elongated or worm holes
- HAZ alteration of microstructure
- Weld metal and HAZ segregation
- Base plate lamination.

The design related discontinuity group might include discontinuities such as stress concentration due to changes in sections, or type of weld joint design.

TYPICAL WELD DEFECTS

Some typical weld defects are shown in Figure 2-8-1. These are some of the commonly detected defects in a typical weld; however their presence, variance and appearance may differ according to the welding process. The soundness of any weld is dependent on the type of defect and its analysis in respect to the end use of the weldments.

The seriousness of any discontinuities should be made on the basis of their shape; the terms used to describe these are planar or three-dimensional. The cracks, laminations, incomplete fusions and inadequate penetrations that have more pronounced stress amplification are the planar type of discontinuity. Inclusions and porosity would be examples of a three-dimensional discontinuity.

While characterizing these discontinuities, the size, acuity (sharpness) and orientation with respect to the principal stress and residual stress directions, the proximity to the weld and the surface of the material are all essential factors to consider.

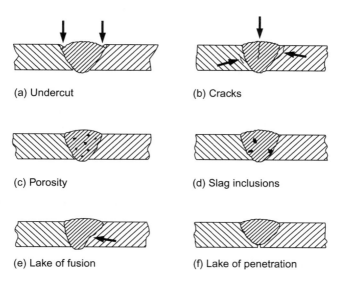

(a) Undercut   (b) Cracks

(c) Porosity   (d) Slag inclusions

(e) Lake of fusion   (f) Lake of penetration

FIGURE 2-8-1 Typical weld defects.

As there are several types of constructions, they generate several types of weld defects and combinations of definitions. These definitions also vary according to the use of the inspection process, and the specification in use. Generally, all these are based on the end use of the weldment. The following is a description of some of the basic types of weld discontinuity.

Porosity

Porosity is the result of gas being entrapped in solidifying weld metal and is generally spherical but may be an elongated void in the weld metal. Uniformly scattered porosity may be scattered throughout single weld passes or throughout several passes of a multi-pass weld.

Faulty welding technique, defective materials or poor weld hygiene generally cause porosity. Cluster porosity is a localized grouping of pores that can result from improper arc initiation or termination. Linear porosity may be aligned along a weld interface, root, or between beads. It is caused by contamination along the boundary.

Piping porosity is elongated and, if exposed to the surface, indicates the presence of severe internal porosity. Porosity has little effect on strength, some effect on ductility, and significant effects on fatigue strength and toughness. External porosity is more injurious than internal porosity because of the stress concentration effects.

Porosity is considered as a spherical and non-planar imperfection, thereby not a serious threat to weld in normal loading and even in fatigue conditions. The tolerance of porosity in such conditions is liberal, and often up to 5% porosity in a standard radiograph length is accepted. However if the weld is required to have good toughness to resist brittle fracture at lower temperatures, especially where the Charpy impact values of the weld metal are required to be 40 J (30 ft-lbf) or higher, the porosity is treated as a planar imperfection.

Inclusions

Slag inclusions are non-metallic particles trapped in the weld metal or at the weld interface. Slag inclusions result from faulty welding technique, improper access to the joint, or both. Sharp notches in joint boundaries or between weld passes promote slag entrapment.

With proper technique, slag inclusions rise to the surface of the molten weld metal. Tungsten inclusions are tungsten particles trapped in weld metal deposited with the GTAW process. Dipping the tungsten electrode in the molten weld-metal or using too high a current that melts the tungsten can cause inclusions.

Incomplete Fusion

Incorrect welding techniques, improper preparations of materials, or wrong joint designs promote incomplete fusion in welds. Insufficient welding current, lack of access to all faces of the joint and insufficient weld joint cleaning are additional causes. This is a stress-concentrating flaw and can in most cases initiate cracks.

Inadequate Joint Penetration

When the actual root penetration of a weld is less than specified, the discontinuity at the root is called inadequate penetration. It may result from insufficient heat input, improper joint design (metal section too thick), incorrect bevel angle, or poor control of the arc.

Some welding procedures for double groove welds require back-gouging of the root of the first weld to expose sound metal before depositing the first pass on the second side to ensure that there is not inadequate joint penetration.

Like incomplete fusion discussed above, incomplete penetration is a stress-concentration point, and cracks can initiate in the un-fused area and propagate as successive beads are deposited. Cyclic loading can initiate catastrophic failures from incomplete penetration.

Undercut

Visible undercut is associated with improper welding techniques or excessive currents, or both. It is a groove running parallel along the toe of the weld, either on the root or the side faces of the weld. The undercuts are of varying depth and area, but it is their relative depth and length that are of importance as acceptable or not acceptable flaws.

Undercut creates a mechanical notch at the weld toe. In addition to the stress raiser caused by the undercut notch, fatigue properties are seriously reduced.

Underfill

Underfill results from the failure to fill the joint with weld metal, as required. It is corrected by adding additional layers of weld metal. If left uncorrected, this becomes a stress concentration point and may subsequently be a cause of the failure.

Overlap

Incorrect welding procedures, wrong welding materials or improper preparation of the base metal causes overlap. It is a surface discontinuity that forms a severe mechanical notch parallel to the weld axis. Fatigue properties are reduced by the presence of the overlap, as they are capable of initiating cracks.

Cracks

In most cases cracks occur when the localized stresses exceed the tensile strength of the material. Cracking is often associated with stress amplification near discontinuities in welds and base metal, or near mechanical notches associated with weldment design.

Hot cracks develop at elevated temperatures during or just after solidification. They propagate between the grains. Cold cracks develop after solidification, as a result of stresses. Cold cracks are often delayed and associated with hydrogen embrittlement. They propagate both between and through grains. Throat cracks run longitudinally in the face of the weld and extend toward the root. Root cracks run longitudinally and originate in the root of the weld. Longitudinal cracks are associated with high welding speeds (such as during SAW) or with high cooling and related restraint.

Transverse cracks are perpendicular to the weld and may propagate from the weld metal into the HAZ and base metal. Transverse cracks are associated with longitudinal shrinkage stresses in weld metal that is embrittled by hydrogen. Crater cracks are formed by improper termination of the welding arc. They are shallow hot cracks.

Toe cracks are cold cracks that initiate normal to the base metal and propagate from the toes of the weld where residual stresses are higher. They result from thermal shrinkage strains acting on embrittled HAZ metal.

Under-bead cracks are cold cracks that form in the HAZ when three conditions are met:

- Hydrogen in solid solution
- Crack-susceptible microstructure
- High residual stresses.

None of the above can be detected by visual inspection and do not normally extend to the surface.

Cracking in any form is an unacceptable discontinuity and is the most detrimental type of welding discontinuity. Cracks must be removed.

Surface Irregularities

Surface pores are caused by improper welding technique such as excessive current, inadequate shielding, or incorrect polarity. They can result in slag entrapment during subsequent welding passes.

Variations in weld surface layers, depressions, variations in weld height or reinforcement, non-uniformity of weld ripples, and other surface irregularities can indicate that improper welding procedures were not followed, or that the welding technique was poor.

BASE METAL DISCONTINUITIES

Base metal properties such as chemical composition, cleanliness, laminations, stringers, surface conditions, and mechanical properties can affect weld quality. Laminations are flat, elongated discontinuities found in the center of wrought products such as plate. They may be too tight to be detected by ultrasonic tests. De-lamination may occur when they are subjected to transverse tensile stresses during welding.

Lamellar tearing is a form of fracture resulting from high stress in the through-thickness direction. Lamellar tears are usually terrace-like separations in the base metal caused by thermally induced shrinkage stresses resulting from welding.

Weld profiles affect the service performance of the joint. Unfavorable surface profiles on internal passes can cause incomplete fusion or slag inclusions in subsequent passes.

DESIGNING WELD JOINTS

A welding engineer will be required to support the activities of the design team and will be called upon to analyze the structural soundness of the members, or evaluate the best ways to eliminate possible flaws in the structure. This requires the welding engineer to be aware of the factors that are responsible for structural stability and soundness, allowable stresses and effect of various loadings, and apply them to the welding part to design the most compatible weldments.

The welding engineer will be required to design weldments to meet specified structural properties. The weld does not stand isolated from other members of the weldment, it has to be an integral part of the entire structure and function accordingly. With the knowledge of the subject discussed in the previous section and chapters it is possible for an educated engineer to address the challenges of structural stability in terms of metallurgical and welding knowledge.

However there are other factors associated with successful designing of weldments. Some of these details are discussed in subsequent chapters and sections of this book.

One of the important parts of designing is a knowledge of structural engineering and analysis of the various stresses associated with any structure. In designing a weld, all principles of structural design have to be evaluated and married to the properties of the material and the weld. This emphasizes the proper selection of material for any specified project. The performance of a structural member is dependent on the properties of the material and the characteristics of the particular section.

The engineers should be able to compute the applied forces on a specific structural member and be able to determine the required strength to resist those forces to prevent the member from failing.

The engineer should take into account and utilize available resources to reduce the cost of the new design, or redesign of the existing structure to upgrade its performance.

The structural integrity of the member should be the prime concern of the designers. For this, the designers should consider the quality of available workmanship, load factors, deterioration from corrosion and damage over the life of the designed structure, safety factors and regulatory requirements.

Basis of Welded Design

Designing for strength and rigidity requires that the structural member must be evaluated to meet the following two parameters in order to meet the service demands:

1. It should not break or yield upon loading, with a factor of safety applied.
2. The loading could be static or dynamic or sometimes the load may be just the weight of the structure itself (called dead load).

In designing, such terms as stress, load, member strain and deformation are commonly used. They also appear in the engineering formulae used by design engineers. Some of these are discussed here in the context of weld design.

In the design process a welding engineer may be required to find out the internal stress of deformation, and the external load. It is normally desirable to keep deflection to the minimum possible. A simple step is to calculate the stiffness of a cantilever beam; the following relationship between applied forces, length of the beam, moment of inertia, and modulus of elasticity computes the amount of vertical deflection at the end of the beam:

$$\Delta = (FL^3)/(3EI) \tag{1}$$

Where:

Δ = The deflection
F = Applied load or force

L = Length of the beam

E = The modulus of elasticity of the material

I = The moment of inertia of the beam section.

In the above equation, the values of E and I should be the largest possible to achieve optimum position that has minimum deflection. For steel, the modulus of elasticity is taken as 30×10^6 psi. I is the moment of inertia of the beam; for this to be large the cross section of the beam should be large enough to limit the deflection to a permissible value. This is a simple condition. There are several out-of-plane forces applied on any member and those are the ones that need to be understood and considered in the design calculations.

There are five basic types of load. When one or more of these loads are applied, they induce stress, in addition to any residual stress contained within the material. These stresses result in strains or movements within the member, the magnitudes of which are governed by the modulus of elasticity of the metal, denoted by letter E. There is always some movement when any load is applied.

We shall now briefly discuss all five types of stress.

Tension

The most common loading on any member is the tensile stress. A straight bar under simple axial tensile force has no tendency to bend. The load causes axial strain and elongation of the bar. The only requirement to prevent it is to have adequate cross-sectional area to withstand the load.

Compression

Compression load causes buckling; an example of this would be a column that is loaded through its center of gravity, under the resulting simple axial stresses. If the column is a slender structure, where the slenderness is defined as the ratio of length to radius of the column, it will start to bow laterally as the load is increased. This would happen at a stress much lower than the yield strength of the material.

Further loading of the column, which is now eccentric to the axis of the load, will cause a bending moment. At this point, if the load is made steady the column will remain stable, due to the combined effect of axial stress and the bending moment. However if the load increases, then a critical point would be reached in the curvature of the column's axial bending, where the column will buckle and fail. From a welding point of view, as the column deflects under load, a bending moment can develop in semi-rigid or rigid welded end connections.

The compressive strength of the column described above depends on its cross-sectional area and the radius of gyration.

To determine the compressive load that a column can support without buckling, the allowable compressive stress is multiplied by the cross-sectional area.

The distance from the neutral axis of the section to an imaginary line in the cross section about which the entire area of the section could be concentrated and still have the same moment of inertia about the neutral axis of the section is the radius of gyration. This indicates the ability of the column to resist buckling. The radius of gyration can be determined by the following equation:

$$r = (I/A)^{0.5} \tag{2}$$

Where:
r = Radius of gyration
I = Moment of inertia about the neutral axis
A = Cross-sectional area of the member

It is important to use the minimal value for the radius of gyration in ratio to the unbraced length of the member. The smaller of the two moments of inertia about axes x-x or y-y is used in the calculation.

Often, industrially published data is used to obtain the values of compressive stress for a given radius of gyration. The American Institute of Steel Construction (AISC) publishes such data for vast numbers of structural steels.

Bending

Bending can occur under uniform or non-uniform loading. Loading within the elastic range results in zero bending stresses along the neutral axis of the member. This stress increases towards the outer fibers of the material.

We know that in a simply supported beam the bending moment decreases along the length as we move toward the end of the beam. This in turn reduces the compressive as well as the tensile stresses that are the reduction of bending stress. In an example of a beam in the shape of an I-section, the bending stress in the flange of the beam will decrease as we move towards the end of the beam. If we consider a short length of the beam, where the tensile forces are F_1 and F_2 applied at these two points, we will note that there is a clear difference between these two tensile forces. A tensile force F is the product of tensile stress (σ_t) and flange cross-sectional area (A). The decrease in the tensile force F in the flange causes a corresponding shearing force between the flange and the web of the beam. This shear force is transmitted by the fillet weld that joins the two (the web and the flange) together. The upper flange that is in compression mode has a similar shear force acting on it.

Another way in which the bending problem manifests itself is called deflection. Deflection is often associated with beams that have larger spans in relation to the depth of the beam. To meet stiffness requirements, the beam depth should be as large as is practical. Where such a large beam depth is not possible, but beam stiffness must be achieved, it is good engineering practice to attach stiffener plates by welding to the web and flange of the beam.

Shear

Shear stresses can be both horizontal and vertical; they also develop diagonal tensile as well as diagonal compressive stresses. The shear capacity of a beam is dependent on its slenderness ratio; this means that the cross-sectional area of the beam is one relevant factor. For the web of a welded beam, if this ratio is less than 260, the vertical shear load is resisted by pure beam shear without lateral buckling to a level of loading well above, where unacceptable deflection will develop. This implies that the design should be based on keeping the shear stress on the gross area of the web below the allowable value of $0.4\sigma_y$ to prevent shear yielding.

In cyclic loading to a level that would initiate web shear buckling, each application of the critical load will cause a breathing action of web panels. This would be due to the plane bending stresses at the toes of the web-to-flange fillet welds and stiffener welds. These cyclic stresses would eventually cause fatigue cracking; the web stresses in these conditions should be limited to the values where shear buckling can be kept out.

In a flange joining the web of a beam fabricated by welding, the shear stress in the weld can be computed using the following equation:

$$W_s = (Vay)/(In) \tag{3}$$

Where:
W_s = Load per unit length of weld
V = External shear force on the member at this location
a = Cross-sectional area of the flange
y = Distance between the center of gravity of the flange and the neutral axis of bending of the whole section
I = Moment of inertia of whole section about the neutral axis of bending
n = Number of welds used to attach the web to the flange.

Torsion

Torsional resistance increases significantly with closed cross-sectional frames. The torsional resistance of a solid rectangular section that has a width which is several times its thickness can be calculated from the following equation:

$$R = bt^3/3 \tag{4}$$

Where:
R = Torsional resistance, (inch^4)
b = Width of the section, (inch)
t = Thickness of the section, (inch)

The total angular twist or rotation of a member can be estimated by the equation given below:

$$\theta = Tl/GR \tag{5}$$

Where:
θ = Angle of the twist (radians)
T = torque, lbf* (in)

l = Length of the member, (inch)
G = Modulus of elasticity in shear, psi (12×10^6 for steel)
R = Total torsional resistance, (inch^4)

The unit angular twist (Φ), is equal to the total angular twist (θ), divided by the length(l), of the member.

Cylindrical or spherical pressure vessels, including hydraulic cylinders, gun barrels, pipes, boilers and tanks are commonly used in industry to carry both liquids and gases under pressure. They can have various shapes and types; for example they may be open ended, cylinders like a pipe, or closed ended cylinders, like ASME pressure vessels, or a sphere. When these vessels are exposed to pressure, the material is subjected to pressure loading, developing stresses in all directions. The stresses resulting from this pressure are functions of the radius of the element under consideration, the shape of the pressure vessel, and the applied pressure.

Calculations similar to those described above are carried out to establish the integrity of cylindrical equipment. These may be a pipeline, or a pressure vessel working under either internal or external (vacuum) pressure. The stresses that work on these fall into a combination of the five types of stresses discussed above.

The internal pressure is often caused by hydrostatic pressure, and in a cylinder this causes stresses in three dimensions:

1. Longitudinal stress (axial) σ_L
2. Radial stress σ_r
3. Hoop stress σ_h

All three stresses are normal stresses, and they need to be computed to meet specific design requirements.

Stresses in Pressure Vessels

The stresses applicable in a pressure vessel are grouped into two types, the thin wall method and the method based on an elasticity solution.

The most common method of analysis is based on a simple mechanics approach and is applicable to 'thin wall' pressure vessels. A thin wall is defined as a vessel whose inner radius (r) to wall thickness (t) ratio is equal to or greater than 10.

The second method is based on an elasticity solution, and is always applicable regardless of the r/t ratio. It can be referred to as the solution for 'thick wall' pressure vessels. Both types of analysis are addressed in various pressure vessel codes like ASME section VIII, BS 1500 and BS 5500, IS 2825 etc. Readers are advised to refer to these codes and standards to adequately familiarize themselves with the details, and to be able to meet specific code requirements.

Pipelines

As stated above pipelines have similar stresses to pressure vessels, however buried lines are subject to another stress in the form of upheaval buckling (UHB), often caused by the soil restraints. If the sturdiness of pipe design is not able to counter the translated effect of thermally induced axial expansion, the resulting expansion and out-of-straightness causes UHB. Should the lateral restraint exceed the vertical uplift restraint – a function of pipeline weight per unit length, pipeline bending stiffness and soil cover – a pipeline may become unstable, and move dramatically in the vertical plane. This phenomenon is described as UHB. If, however, the vertical uplift restraint exceeds the lateral restraint, the resultant pipeline expansion would feed into buckling sites in the lateral plane.

The occurrence of UHB is a result of pipeline out-of-straightness, which may be introduced by trench bed imperfections, lack of straightness of the pipe, or misalignments of the girth weld. Predicting possible UHB is key to the proper design of a buried pipeline, and can be complex because of uncertainties in the soil properties and pipeline vertical profile, as well as the specific project requirements. If not predicted and mitigated, UHB can result in significant damage to the pipeline.

FIGURE 2-8-2 Transportation of fabricated items through waterways to the offshore staging points.

FIGURE 2-8-3   Typical construction site – Note the main line valves welded to the pipeline.

FIGURE 2.4.5

NON-DESTRUCTIVE TESTING

Chapter 1

Introduction

A welding engineer designs weldments to meet the rigors of the design conditions. Other than design simulations, and mathematical modeling where this is possible, there are only two ways to factually determine if the design will really work. One of these is to put every designed product into a real life test, but this has some very serious implications:

- Real life testing may require huge test facilities, which would be prohibitively expensive
- The real life test may destroy the weldments, and so it may not be tested to full load.

A second option is to put the weldment through a testing regime by computer simulations and mathematical modeling. This can be complemented by selective tests of material properties during the production process, involving inspection of materials and welds by non-destructive testing methods. This would cover most aspects of the weldment. In some cases, such as pressure vessels, piping and pipelines, an additional pressure test is also carried out to prove the design.

The general term non-destructive testing (NDT) or sometimes non-destructive examination (NDE) is used to identify all those inspection methods that permit evaluation of materials and welds without destroying them.

This section of the book will introduce various non-destructive testing methods and their fundamental principles and typical applications in pre-construction, during construction and post-construction, together with integrity management of structures, plant and equipment.

The common elements of nearly all NDT methods are:

(a) A source of probing energy
(b) A test specimen that is appropriate for the energy source being used, so that the discontinuities may be detected
(c) A detector that can accurately measure the distribution and changing energy
(d) A technique for recording or displaying the information received from detector
(e) A trained and qualified operator to inspect and interpret the feedback from the detector unit.

Applied Welding Engineering: Processes, Codes and Standards.

The fundamental basis of selecting a particular type of NDT method depends on the available energy, the type of material being tested, the objective of inspection and the degree of accuracy required.

The ability of engineers to take a sound decision to use a single NDT method or a group of NDT methods is affected not only by their knowledge of the various tests available, but also by their knowledge of the limits and advantages of the respective test, and their ability to use them to complement interdisciplinary limitations. Engineers can design a testing protocol more effectively by using their knowledge of the primary and secondary manu-facturing processes involved in fabrication and production, their sequence, including of course a sound knowledge of processes like welding, forging, casting, machining etc. This will be further aided by an in-depth knowledge of the heating and cooling cycles used on various materials.

A case study is attached in an additional section, where real life, practical applications of NDT methods are discussed. Typical defects that are encoun-tered in a material after processing or during maintenance and welding or due to corrosion are discussed. The reason why a certain NDT method is more suitable to detect a certain type of defect is also explained.

We shall discuss the test methods listed in Table 3-1-1 below.

TABLE 3-1-1

NDT Methods	Name of the Test Method	Accepted Abbreviations
1	Visual Inspection	VT
2	Radiography	RT
3	Magnetic Particle Testing	MT
4	Penetrant Testing	PT
5	Ultrasonic Testing	UT
6	Eddy Current Testing	ET
7	Acoustic Emission Testing	AET
8	Ferrite Testing	FT
9	Leak Testing	LT
10	Proof Testing	PRT

Visual Inspection (VT)

Visual inspection (VT) is an important part of the quality control system. For several machines, materials and welds it is the primary means of inspection. It is the simplest form of inspection, sometimes performed consciously as part of a structured inspection program, but on other occasions it is performed as an unconscious activity as part of day-to-day work. Sometimes visual inspection is the only inspection given to materials or welds, but most of the time it is a prerequisite before the application of any other test method. In most cases it acts as the first line of filtering for faulty or defective materials and welds:

- The inspection of the jet engines upon arrival from flight involves an extensive regimen of visual inspections
- Regular inspection of a ship's engine room, decks and other equipment is a preventive measure to identify and avoid major failures and costly repairs
- A daily walk around your car before driving is a visual inspection to look at the tyre tread, and identify any other possible damage that might become a cause of serious car problems.

The above are just a few examples of visual inspections which are intended to identify, pre-empt and prevent possible catastrophic failures. Some organized inspection routines have very structured systems, with forms and questionnaires to be filled in and associated responsibilities, and others are less structured, depending on the associated (potential) seriousness of the resulting failure, in terms of both life and cost.

Visual inspection in terms of welding assumes some specialized knowledge, training and experience. Depending on the end objective of the inspection or the design of the weldments, some demands on the attributes, training, knowledge and experience of the inspector will differ. Training as an inspector will generally involve classroom education and written examinations, alongside some real, hands-on inspection of welds. Such certification is carried out by several national certification bodies like The Welding Institute's welding inspector certification, www.twi.co.uk, The American Welding Society welding inspector certifications, www.aws.org, or The Canadian Welding Bureau, eng.cwbgroup.org. These certifying bodies also determine that the inspector has correct vision acuity for distance and color vision. Generally the ability to read J-1 letters on a Jaeger standard chart at a distance of 12 inches is deemed to be an acceptable near-vision level.

Visual inspection requires that the area to be inspected is fully illuminated, so that the inspector is able to see the details of the weldment that is to be inspected. A minimum illumination of 1,000 Lux (100 fc) on the surface is considered an adequate light for visual inspection.

ADVANTAGES OF VISUAL INSPECTION

The advantages of this test are as follows:

1. It is easy to apply.
2. It is quick to apply.
3. It is relatively inexpensive.
4. Generally no special equipment is needed.

There is certain physical ability and knowledge expected of an inspector, to be efficient in the job.

As stated above the inspector should have good eyesight, and should be able to move around and reach difficult to reach positions of the material or weld being inspected. The inspector is expected to have good knowledge of the machinery, equipment and specifications, workmanship standards, and should be familiar with shop practices, that they are required to inspect.

For conducting an effective visual inspection the inspector may need the following additional tools:

1. A source of light if natural light is not adequate.
2. Internal inspection of a restricted location can be done with the help of a boroscope, a reflecting glass with added light comes handy in some restricted inspection locations, like jet engine's turbine blades, corner welds, welds inside a small bore tubing etc.
3. A low powered magnifying glass may be required in some specific cases.
4. Some hand tools and in some cases some specialized tools of the trade may be required.

Welding inspection poses a different set of challenges. For inspecting welds during construction, the following tools may be required:

- Weld profile gauge
- Fillet gauge
- Gap gauge

- Bevel protector
- A ruler, Vernier gauge or a tape measure of suitable scale.

Inspection of welding can be divided into three stages as listed in the following table. Table 3-2-1 lists the general inspection scope for all three stages.

The general acceptance criteria for a weld profile are given in Table 3-2-2. Various specifications/codes have their own acceptance criteria, and may differ significantly from the limits given in this table. An example of this is the acceptance criteria for welds in deep-sea Steel Contrary Risers (SCRs), where an acceptable profile is limited to 'ground flush' to parent metal. The alignment given in Table 3-2-3 is similarly stringent, often limited within the range of 0.001 to 0.004 inches.

TABLE 3-2-1 The Three Stages of Welding and What to Look For at Each Stage

Welding Inspection Stages	Inspection Points Addressed in These Stages
Prior to welding	Is material being welded as per the specification and as per fabrication drawing?
	Is the joint preparation as per the approved welding procedure and drawing?
	Clearances, dimensions, backing strips, backing material or gas filler metal.
	Alignment, fit-up, distortion control etc.
	Verification of WPS/PQRs and cleanliness.
During welding	Welding process and conditions.
	Filler metal flux or shielding gas.
	Preheat and interpass temperature and control.
	Distortion control.
	Interpass chipping, grinding, gouging, and cleaning.
	Inspection intervals.
Post-welding	Dimensional accuracy.
	Conformity to drawing and specifications.
	Acceptability with regards to appearances (including weld spatter, roughness, undercuts, overlaps, cracks, open pores, craters etc.).
	Post-weld heat treatment time and temperature.
	Release for other NDE methods, as applicable.

TABLE 3.2.2 Weld Reinforcement Acceptability Level

Material nominal thickness (mm)	Max reinforcement (mm)	Max reinforcement where more stress is expected (mm)
Less than 2.4	2.4	0.8
Over 2.4 to 4.8	3.2	1.6
Over 4.8 to 13	4.0	2.4
Over 13 to 25	4.8	2.4
Over 25 to 51	5.0	3.2
Over 51 to 76	6.0	4
Over 76 to 102	6.0	6
Over 102 to 127	6.0	6
Over 127	8.0	8

Radiography

Other than very basic fabrications, most weldments require more than just visual inspection. The design may demand that the health of the weld be evaluated in more detail. Such demands may push for additional methods of inspection. Radiography is one of those methods, in which a two-dimensional picture of the three dimensions of weld can be produced, revealing its internal health.

This non-destructive testing method utilizes radiation to penetrate the material and records images on a variety of recording devices such as film, photosensitive paper, or fluorescent screen. When the radiation passes through a metal, some of it will be absorbed by dense material, some will be scattered and some will be transmitted through the less dense metal. The resultant radiation energy will be varied and will be recorded accordingly on the viewing media.

For ease of understanding the weld inspection, we will focus our discussion on the most common recording medium, film.

The basic process of radiography involves two stages: the first is obtaining a good, readable radiograph, and the second is the ability to interpret whatever

TABLE 3-3-1

Points	Essentials for taking good radiograph	What is required?
1	A source of radiation	Gamma or X-ray
2	An object to be rediographed	Material or weld
3	Film enclosed in a lightproof holder	Suitable quality of film
4	Producing an exposure in the most advantageous manner	A trained person
5	Processing the film	A chemical processing bath.
Points	Requirement for interpretation of film	General requirements
1	Interpretation	A trained and experienced person

the radiograph presents. The entire process revolves around these two very basic fundamentals. As we delve deeper in the process, the two stages become more technical and it is limited only by depth of the knowledge the reader wants to acquire.

The requirements for achieving the two basic goals are varied and sometimes very divergent routes are taken. Table 3-3-1 summarizes the essential requirements for accomplishing a good radiograph. In the subsequent paragraphs, each of these will be discussed in more detail.

SOURCE OF RADIATION

The heart of radiography is the radiation source. Electromagnetic radiation energy has penetrating properties that are related to its energy potential or wavelength. Another unique quality of radiation is its ability to ionize elements. These two unique properties are exploited to get good radiographs.

These radiation sources have extremely short wavelengths, of about 1/10 000th that of visible light. This enables the source to penetrate the material. The two types of radiation source commonly used for industrial radiography are:

- X-rays
- Radioactive isotopes – gamma rays.

For radiographic purposes, both x-ray and gamma ray radiation behave similarly. Both comprise a high energy, short wavelength portion of the electromagnetic spectrum. Throughout the spectrum, x-rays and gamma rays have the same characteristics, and x-rays and gamma rays of the same wavelength have identical properties. The characteristics of this type of radiation are listed below.

1. They have no electrical charge and no rest mass.
2. In free space they travel in straight lines at the velocity of light.

3. They are electromagnetic with energy inversely proportional to their wavelength.
4. They can penetrate matter, with the depth of penetration being dependent on the wavelength of the radiation and the nature of the matter being penetrated.
5. They are invisible and incapable of detection by any of the senses.
6. They can ionize matter.
7. They are absorbed by matter. The degree of absorption is a function of the matter density, its thickness and the wavelength of the radiation.
8. They are scattered by matter, the amount of scatter is a function of matter density and wavelength of radiation.
9. They can expose film by ionization.
10. They can produce fluorescence in certain material.

X-RAYS

We shall briefly discuss x-rays and how they are generated for industrial applications.

X-ray radiation is obtained from a machine that contains a tube consisting of an electron source, a target for electrons to strike, and a means of speeding the electrons in the desired direction.

- Electron Source
  All matter is composed in part of negatively charged electrons. When a suitable material is heated, some of its electrons become agitated and escape the material as free electrons. These free electrons surround the material as an electron cloud. In an x-ray tube the source of electrons is known as the cathode. A coil of wire called a filament is contained in the cathode and it functions as the electron emitter. When a voltage is applied across the filament, the resultant current flow heats the filament to its electron emission temperature.
- Electron Target
  X-rays are generated whenever high-velocity electrons collide with any form of matter, whether it be solid, liquid, or gas. Since the atomic number of an element indicates its density, the higher the atomic number of the chosen target material the greater the efficiency of x-ray generation. In other words, the greater the density of the material, the greater the number of x-ray generating collisions. In practical applications, a solid material of high atomic number, usually tungsten, serves as the target. As shown in Figure 3-3-1 the target is contained in the x-ray tube on the anode.
- Electron Acceleration
  The electrons emitted at the cathode of an x-ray tube are negatively charged. They are repelled by negatively charged objects and are attracted to positively charged objects. The anode of the x-ray tube is positively charged so it attracts the negatively charged electrons from the cathode, thus electrons are speeded from cathode to anode.

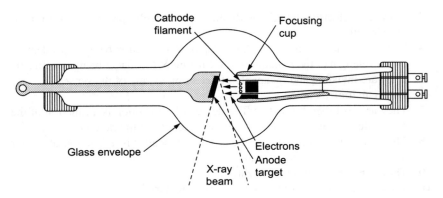

FIGURE 3-3-1   Typical basic x-ray tube.

- Intensity

  The amount of x-rays created by electrons striking the target is one measure of the 'intensity' of the x-ray beam. Intensity is therefore dependent
  upon the amount of electrons available at the x-ray tube cathode. If all
  other variables were held constant, an increase in cathode temperature
  would cause emission of more electrons and increase the intensity of the
  x-ray beam. To a lesser degree an increase of positive voltage applied to
  the anode will increase the beam intensity. Thus the intensity of the x-ray
  beam is almost directly proportional to the flow of electrons through the
  tube. The output rating of an x-ray machine is often expressed in milliam-
  peres of current flow. This 'direct proportion' establishes tube current as
  one of the exposure constants of x-ray radiography.

- Inverse Square Law

  The intensity of an x-ray beam varies inversely with the square of the distance from the radiation source. X-rays, like visible light rays, diverge
  from the radiation source and cover increasingly large areas as the distance
  from the source is increased. This is an important law that helps compute
  radiography exposure and safety procedures. Mathematically the inverse
  square law is expressed as:

$$I_1 / I = D^2 / D_1^2$$

Where:

  I = is the x-ray beam intensity at distance D and
  I_1 = is the x-ray beam intensity at distance D_1

- X-Ray Quality Characteristics

  Radiation from an x-ray tube consists of previously mentioned characteristics and continuous rays. The characteristic rays are of small energy
  content and specific wavelengths as determined by the target material. The

spectrum of continuous rays covers a wide band of wavelengths and is of generally higher content. The continuous rays are of most use in radiography. Since the wavelength of an x-ray is partially determined by the energy (velocity) of the electron whose collision with the target produced it, an increase in applied voltage will produce x-rays of shorter wavelength (higher energy). An increase in applied voltage also increases the intensity (quality) of x-rays. But of more importance to radiography is the generation of higher energies with greater penetration power.

High-energy rays (short wavelength) x-rays are known as hard x-rays, whereas low-energy (longer wavelength) x-rays are known as soft x-rays.

EFFECT OF KV AND MA

To understand the interaction of x-rays with matter, it is necessary to consider the properties of matter that make this interaction possible. Matter is composed of tiny particles called atoms. A substance composed only of identical atoms is called an element, and substances composed of two or more elements are called compounds. Atoms are themselves made up of even smaller particles, as listed and defined in the table below.

TABLE 3-3-2 Effect of Kv and MA

	Low MA	High MA
Low Kv	Low intensity soft x-rays	High intensity soft x-rays
High Kv	Low intensity hard x-rays	High intensity hard x-rays

TABLE 3-3-3 Fundamental Particles

Particle	Description
Proton	A particle carrying a unit positive electrical charge. Its mass is approximately one atomic mass unit.
Neutron	A particle, electrically neutral. It has approximately the same mass as the proton.
Electron	A particle carrying a unit negative charge. Its mass is 1/1840 atomic mass unit.*
Positron	A particle carrying a unit positive electrical charge. Its mass is the same as an electron.

*The atomic mass unit (AMU) is 1/12 the mass of the carbon-12 atom.

SCATTER RADIATION

In passing through matter, x-rays lose energy to atoms by three ionization processes:

1. Photoelectric absorption.
2. Compton scattering.
3. Pair production.

All three processes liberate electrons that move with different velocities in various directions. As we know that x-rays are generated when free electrons collide with matter, it follows that these x-rays passing through matter cause the generation of secondary x-rays. These secondary x-rays are minor components of what is called scatter radiation.

The major components of scatter are the low-energy rays represented by photons- weakened in the Compton scatter process. Scatter radiation has a uniformly low-level-energy content and is of random direction.

In radiography the scatter occurs in three ways; internal, side and back scatter. Figure 3-3-2 demonstrates the practical aspect of eradiation scatter as it affects the quality of radiographs.

Internal scatter is the scattering that occurs within the specimen being radiographed. It is reasonably uniform throughout a specimen of a single

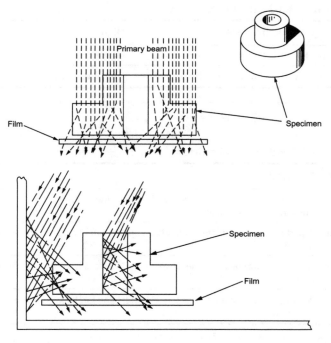

FIGURE 3-3-2   Graphical representation of internal and side scatter.

thickness; it blurs the image and reduces its definition. The loss of definition generally occurs around edges and any sharp corners like a bored-hole, step-downs, grooves and recesses cut in the specimen.

Side scatter is the scattering from walls of the objects within the vicinity of the specimen or from portions of the specimen itself that cause the rays to re-enter the specimen. Side scatters also blur and obscure the definition of the edges.

Backscatter, as the name suggests, is the scattering of rays from surfaces or objects beneath or behind the specimen. Backscatter has a similar effect on the edges of the specimen.

X-RAY EQUIPMENT

These machines are available in a variety of ranges. They range from as low as 50 kV to up to 30 million electron volts output capacity. The general concept of these machines is discussed in the following paragraphs to provide a basic understanding of them.

These machines require electrical energy to produce the required radiation and, because of this, their portability is limited. The portability is also affected as a typical x-ray tube consists of the following components, making it bulkier, not very easy to make it a movable machine. An x-ray machine consists of the following parts:

1. Tube envelope
2. Filament heating
3. Anode
4. Focal point

All the above are depicted in Figure 3-3-1.

It is important to note these radiographic terminologies, which will be repeated as we discuss the quality of radiographs in the rest of this chapter.

The sharpness of a radiograph is particularly dependent on the size of the radiation source, also called the focal spot. The electron beam in most x-ray tubes is focused so that a rectangular area of the target is bombarded by the beam. The anode target is set at an angle, and the projected size of the bombardment area as received by the specimen is smaller than the actual focal spot. The projected area of the electron beam is the effective focal spot. This property allows reduction of the focal spot as desired; in theory the limit of reduction is unrestrained and often called pinpoint reduction, but in practice the size of the focal spot is limited by the generation of heat that can destroy the target.

Power Sources

Transformers are often the source of electric power; there are various types of iron core transformers which produce voltages of up to 400 kvp. They produce self-rectified circuits of half wave or full wave rectification, voltage doublers, and constant-potential. They may be mounted in the tube head tank units or separately housed.

Resonant transformers often range from 250 to 4,000 kvp. These transformers produce high voltage from low input. This type of transformer allows more compact design, as the x-ray tube can be mounted in the central axis of the transformer.

Electrostatic generators operate in the range of 500–6,000 kvp. Two motor driven pulleys drive a (non-conducting) charging belt. Electrons from the charging point pass to the belt and are transferred to the corona cap at the corona point. The accumulated high voltage from the corona cap helps to accelerate the beam of electrons emitted by the filament. The equipotential plates distribute high voltage evenly along the length of the tube. The generator is enclosed in a pressurized gas tight chamber to minimize high voltage leakage.

Linear accelerators utilize radio frequency energy in a tuned waveguide to produce an induced field, which is directly related to the length of the waveguide section and the radio frequency. Electrons injected into the guide are accelerated towards the target by the action of the constantly changing induction field.

This should accelerate the electrons to velocities which reach the speed of light, and so produce extremely short wavelength and high frequency x-rays, but at this level of application, there is a disconnect between the theory and practice of the process, at least at this level of application. Practically, the length of the linear accelerator that is required to obtain electron velocities equivalent to those used in industrial radiography is about 6 feet.

Betatron accelerators use magnetic circular induction to accelerate electrons, on the same principle that a transformer works. Alternating current is applied to the primary coil to produce strong variation in the magnetic field in the core of the secondary. The magnets strengthen the magnetic field. X-rays of extremely short wavelength and great penetration power are produced.

Control Panel

The control panels of x-ray machines are designed to allow the operators to control the generation of x-rays to produce exposures. It also serves as a protective layer for the equipment.

As we have seen, there are various types of equipment, and therefore control panels will differ from each other, but a general overview can be made of the different components of a typical panel. Note that some of the features discussed may be differently arranged or may not be part of a specific machine as they all differ in design and use.

1. Line voltage selector switch. This permits the equipment operators to connect to different input sources.
2. Line voltage control. This permits adjustment of line voltage to exact values.

3. Line voltage meter. A voltmeter indicating the line voltage use in conjunction with the line voltage control.
4. High-voltage control. Permits adjustment of voltage applied across the tube.
5. High voltage meter. A voltmeter, generally calibrated in kilovolts, used in conjunction with the high-voltage control.
6. Tube-current control. This permits adjustment of tube current to exact values.
7. Tube-current meter. An ammeter, usually calibrated in milli-amperes, used in conjunction with tube current control.
8. Exposure time. A synchronous timing device used to time exposures.
9. Power ON-OFF switch. Controls the application of power to the equipment. It usually applies power to the tube filament only.
10. Power ON indicator light. Visual indication that the equipment is energized.
11. High-voltage ON-OFF Switch. Controls the application of power to the tube anode.
12. High-voltage ON indicator light. Visual indication that the equipment is fully energized and x-rays are being generated.
13. Cooling ON indicator light. Visual indication that the cooling system is functioning.
14. Focal-spot selector control. Used with tubes having two focal spots; permits selection of the desired size of focal spot.

GAMMA RAYS

Gamma rays are produced by the nuclei of isotopes undergoing disintegration because of their basic instability. Isotopes are varieties of same chemical elements but have different atomic weights. A parent element and its isotopes both have an identical number of protons in their nuclei, but a different number of neutrons. Among the known elements there are more than 800 isotopes, of these about 500 are radioactive. The wavelength and intensity of gamma waves are determined by the characteristics of the source isotopes and cannot be controlled or changed.

Every element with an atomic number greater than 82 has a nucleus that has the potential to disintegrate because of its inherent instability. Radium is the best known of such elements; it is used as a natural radioactive source. Radium and its daughter products release energy in the form of alpha rays (α rays), which are helium nuclei, consisting of two protons and two neutrons with a double positive charge. Beta rays (β rays) are negatively charged particles with a mass and charge equal in magnitude to those of an electron, and finally gamma rays (γ rays) are short wavelength electromagnetic radiation of nuclear origin.

Both alpha and beta particles exhibit relatively negligible penetrating powers. Hence, it is the gamma rays (γ rays) that are of use to radiographers.

Artificial Sources

The availability of artificial radiation sources has made radiography much easier. There are two sources of man-made radioactive source. The first option is an atomic reactor, which operates by fission of uranium 235, producing several applicable isotopes. Cesium 137 is a byproduct of nuclear fission.

The second option is the creation of isotopes by bombarding certain elements with neutrons. In this process the nuclei of the bombarded element are made unstable or radioactive as they capture neutrons. In this group are the isotopes of cobalt 60, thulium 170 and iridium 192. The numerical designators denote the mass number of the isotope, which distinguishes it from the parent isotope or other isotopes of the same element. Artificial isotopes emit alpha rays (α rays), beta rays (β rays) and gamma ray (γ rays), similarly to natural isotopes.

Gamma ray intensity is measured in roentgens per hour at one foot (rhf) distance. This is a measure of radiation emission over a given period of time at a fixed distance. The activity of a gamma source determines the intensity of its radiation. The activity of an artificial radio-isotope source is determined by effectiveness of the neutron bombardment that created the isotopes. The measure of activity is the Curie – which is 3.7×10^{10} disintegrations per second.

The specific activity of a given gamma ray source is defined as the degree of concentration of radioactive material within that source. It is expressed in Curies per gram, or Curies per cubic centimeter. Two isotope sources of the same material with the same activity, and a different specific activity will have different dimensions. The source with greater specific activity will be the smaller of the two.

The specific activity is an important measure for radiography. It is well known that a smaller source results in greater sharpness of the image on the film. This is shown in Figure 3-3-3, where the size of the source is one factor determining the quality of the radiograph.

Half-Life

Half-life is the amount of time required for an isotope to disintegrate (decay) to one-half of its initial strength. It is specific to the isotope element it comes from. This information is useful as a measure of activity in relation to time. A dated decay curve of each isotope is provided on procurement of a radio-isotope source. A typical decay chart of iridium-192 is shown in Figure 3-3-4. Table 3-3-4 gives the characteristics of several isotopes; it includes the half-life, dose rate in Curie/ft/hour and other safety-related information.

The inverse square law in which 'the intensity of x-rays varies inversely with the square of distance from the radiation source' is also applicable to gamma rays.

Radiation from a gamma source has a wavelength (energy) dependent on the nature of the source. Each of the commonly used radio-isotopes has specific uses, due to their fixed gamma ray energy characteristics.

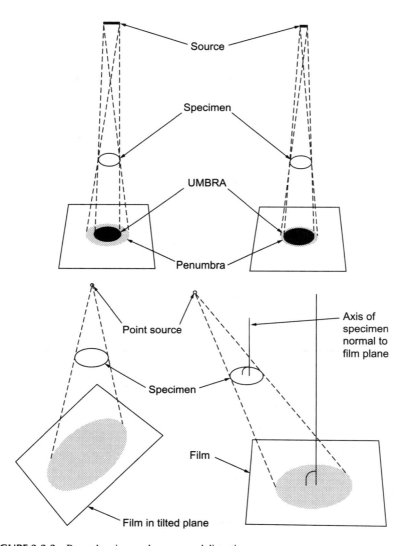

FIGURE 3-3-3   Penumbra, image sharpness and distortions.

Table 3-3-5 gives a list of industrial isotopes and the gamma-ray energy each of them emit.

FILM

Image capture is another important aspect of radiography. The most common method in use is to capture images on film. Radiographic film consists of a thin, transparent plastic sheet coated on one or both sides with an emulsion of

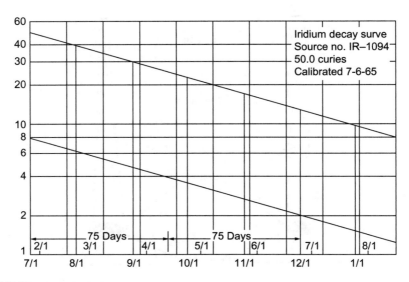

FIGURE 3-3-4   Typical gamma-ray source decay curve.

TABLE 3-3-4 Isotope Characteristics

Isotope	Cobalt-60	Iridium-192	Thulium-170	Cesium-137
Half life	5.3 years	75 days	130 days	30 years
Chemical form	Co	Ir	Tm_2O_3	CsCl
Dose rate at 1 ft per curie	14.4 (r/Hour)	5.9 (r/Hour)	0.032* (r/Hour)	4.2 (r/Hour)
Practical Sources				
Curies	20	50	50	75
RHM	27	27	0.1	30
Approx Diameter	3 mm	3 mm	3 mm	10 mm

gelatin approximately 0.001 inch thick, which contains very fine grains of sil-ver bromide. When this film is exposed to x-rays, gamma rays or visible light, the silver bromide crystals undergo a reaction that makes them more suscepti-ble to the chemical process (called developing). This process converts them to black metallic silver. In other words exposure to radiation rays creates a latent image on the film, and further chemical processing makes the image visible.

The usefulness of any radiograph depends on its impact on human eyes. A good interpretation depends on the contrast and definition detectable by eye. When a radiograph is viewed for interpretation, the details of the image should

TABLE 3-3-5 Gamma Ray Energy

Isotope	Gamma Ray Energy (MEV)
Cobalt 60	1.33, 1.77
Iridium 192	0.31, 0.47, 0.60
Thulium 170	0.084, 0.052
Cesium 137	0.66

be clear and understandable. This depends on the amount of the light passing through the processed film. The areas of high density, appearing dark gray, will be viewed against areas of light density, appearing as various shades of light gray. The density difference between the two is called the contrast of the film. The other attributes that affect the quality of film are the sharpness of the film image – also called the definition of the film.

Film is further discussed in the paragraphs covering the quality of radiographs.

There are several factors that contribute to a successful radiograph, including the geometry of the object being radiographed, the quality of film, the exposure time and processing and viewing conditions.

RADIOGRAPHIC EXPOSURE TECHNIQUES

Radiography is carried out based on the geometrical shape of the material and the required details of the image. The best possible configuration is chosen and the film placement and location of radiation source is determined. In this respect the configuration of cylindrical objects like pipeline and offshore structural members are more critical. The radiographer must ask the following questions and, based on the answers, consider the best possible exposure technique to use:

- Which arrangement will provide the optimum image quality and show the area of interest (weld) coverage?
- Which arrangement will provide the best view of those discontinuities most likely to be present within the material, and of the area of interest?
- Will a multiple film exposure technique be required for full coverage?
- Can a panoramic-type exposure be used?
- Which arrangement will require the least amount of exposure time?
- Can the exposure be made safely?

There are various exposure techniques used to produce the best possible image. Figure 3-3-5 below shows typical exposure arrangements. We will discuss a few of the most common techniques in the following paragraphs. They are commonly used in most plate and tubular construction, including piping, pressure vessels, boilers, and structural fabrications and pipelines.

Single Wall Single Image (SWSI)

For a plate (sketch A) the arrangement is the simplest; it has good exposure geometry and proper positioning of radiographic tools is possible. It is the most practical set-up. The figure shows a plate butt-weld (Object) and a film placed below it. The radiation source is placed opposite the film, keeping the object between the film and the source.

Exposure E is a similar arrangement in a pipe. The film is placed in the pipe close to the weld. The source is outside the pipe at a distance suitable to make a good exposure. Both these techniques are examples of single wall single image viewing.

Panoramic Technique

Sketch F describes the panoramic viewing technique. This is a variant of the SWSI technique. The technique is used for cylindrical objects of relatively large diameter (greater than 90 mm (3.5 inch)). The film(s) are placed on the outer surface close to the weld (material), and a source is placed in the pipe at its center, allowing the radiation to travel equal distances to all parts of the pipe's circular cross section. This allows full girth or the pipe to be exposed all at the same time.

All the above techniques produce an image that has radiation penetrated through a single wall of the material, producing the simplest image of the weld on the film.

Double Wall Single Image (DWSI)

In contrast to the above technique, in sketch B, the radiation source is positioned on the outside of the pipe wall, with the film placed on opposite side. In this set up, the first wall closest to the source is considered nonexistent and the source is offset. This is the double wall single image viewing technique.

Double Wall Double Image (DWDI)

When the pipe diameter is relatively small – that is less than a nominal diameter of 38 mm (1.5 inch), then the entire circumferential weld is exposed straight through both walls, as is shown in sketch C. Images of both walls are projected onto the same film area. The determination of defect location and depth is difficult in this double wall single image viewing. However, if the wall thickness and the diameter permit the source can be offset, an elliptical image

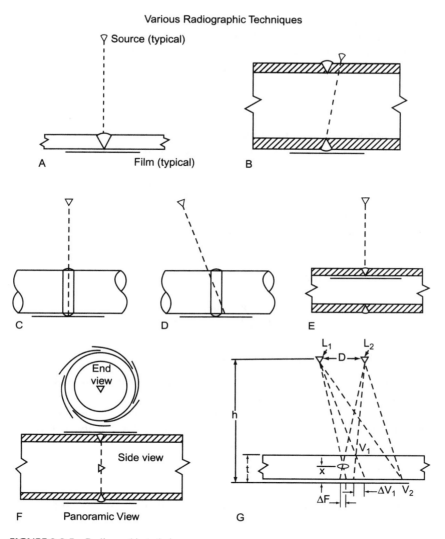

FIGURE 3-3-5   Radiographic techniques.

of the weld is obtained, as is shown in sketch D. An experienced interpreter will be able make best judgment of type and location of any discontinuity that is viewed in the film.

Radiography presents a two-dimensional image of the object, and so it is not possible to determine the depth of any discontinuity. To meet this challenge a technique is used with two exposures (see sketch G). These two exposures are taken in the same setup, on two films with an offset (D) in the location of the source, while maintaining the focus to film distance (h) a constant. The resulting images provide parallax to reveal the third dimension. With fixed lead markers V_1 and V_2 on the source and film sides, the shadow of flaw (F) and marker V_1 will change position to ΔF and ΔV_1. If the thickness of the material being tested is t then the distance of the flaw above the film is calculated as:

$$\text{Distance of the flaw from film } (X) = t * \Delta F / \Delta V_1$$

It is important to note that this calculation assumes that the image of the bottom marker V_2 remains essentially stationary with respect to the film. This may not always be possible; for example if the film cassette or film holder is not in contact with the bottom surface of the test object or larger source shifts are used. In these instances, the location of the flaw may be computed by the following formula:

$$\text{Distance of the flaw from film } (X) = h * \Delta F / D + \Delta F$$

Where:

X = Distance of flaw above film
h = Focus to film distance
ΔF = Change of position of flaw image
D = Distance source has shifted from L_1 to L_2

RADIOGRAPHIC IMAGE QUALITY

Practical radiography is a relatively simple process, but it involves a large number of variables and it requires well organized techniques to obtain consistent quality. The radiographic image must provide useful information regarding the internal soundness of the specimen. Image quality is governed by two categories of variables which control:

1. Radiographic contrast
2. Radiographic definition

Both these general categories can be further divided. We shall briefly discuss them here.

Radiographic Contrast

In a radiograph of a material that has varying thickness, there is different absorption of radiation along the specimen. This results in different film densities, in which the thicker sections appear lighter that the thinner sections on the film. This difference in darkness of the image from one area to another is called 'radiographic contrast'. Any shadow or details within the image are visible by means of the contrast between it and the background. It may be noted that up to a certain degree, the greater the degree of contrast or density differences in the radiograph, the more readily various images will stand out. The radiographic contrast of concern here is the result of the combination of two additional contrast components; subject contrast and film contrast.

Subject Contrast

Subject contrast is caused by the range of absorption of radiation by the specimen being radiographed. This is affected by (a) the mass of the subject, including the atomic number and the thickness, and (b) the penetrating power of the radiation source at the wavelength of the radiation used. The subject contrast is also affected by scattering of the radiation. We have discussed types of radiation scatter earlier in this chapter.

Film Contrast

Film contrast is caused by the characteristics of the film used to record the image. Recording processes, depending upon the film type and related variables, can amplify the difference in film densities created by subject contrast. This is called 'process contrast amplification'. Film emulsion is manufactured to give different degrees of film contrast. They can also affect the speed of the film and its level of graininess.

The principles of film contrast are best understood by looking at the film characteristic curve, also called the H&D curve, and named after Hurter and Driffield who devised it.

The curve relates to the degree of darkness (density) to the logarithm of radiation exposure. The radiation exposure (R) of the film is defined as the

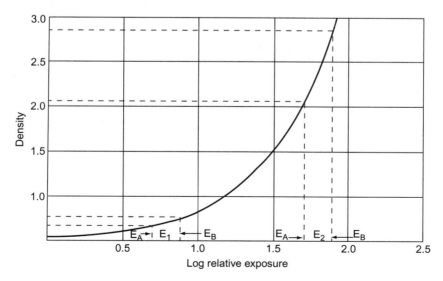

FIGURE 3-3-6   Film characteristic curve.

product of the radiation intensity (I) exposing film and the time (t) of exposure duration. The relationship is expressed thus;

$$R = I * t \tag{1}$$

The exposure value is expressed as a logarithm (usually base 10) for two reasons; one being the convenience of comparing the otherwise very long scale of exposure units, and second, a change in exposure (ΔR) on logarithmic scale has a constant spread throughout the scale, whereas the same ΔR would change in spread along an arithmetic scale. A constant exposure difference is the basic tool that renders a characteristic curve useful.

Film density (D), or optical transmission density, is defined as:

$$D = \log_{10} I_0 / I_t$$

Where:

D = Film density
I_0 = Intensity of light incident on film
I_t = Intensity of light transmitted through the film (as seen by the eyes)

A film density of 1.0 H&D units means that 1 out of 10 parts of light will reach the eye after being transmitted through the film, or:

$$D = \text{Log}_{10} \ 10 \ \text{units}/1 \ \text{unit} = 1.0 \ \text{H\&D density}$$

The shape (slope) of the curve at any particular film density is not only dependent upon the type of film selected, but is also influenced by the degree of development and the type of intensification screens used.

Thus the shape or the slope of the characteristic curve and the density of the film are two factors that define the film contrast.

RADIOGRAPHIC DEFINITION

The definition of a radiographic image is as important as the radiographic contrast discussed above. Definition concerns the sharpness of the image. 'Sharpness' means the degree of abruptness of the transition from one density to another. The more abrupt this transition, the greater the ease in identifying or defining the image. It is easier to discern details in a sharp radiograph than in an out-of-focus one.

The two components of radiographic definition are exposure geometry and film graininess, and both are briefly discussed below.

Exposure Geometry

The sharpness aspect of the exposure geometry is shown in Figure 3-3-3. The penumbral shadow is affected by:

(a) source to film distance
(b) source size
(c) specimen to film distance.

TABLE 3-3-6

Source Size/Focal Spot	Typical Diameter/Area
Gamma Ray Sources	1.6 mm to 13 mm
X-Ray	Varies from 6 to 8 mm² to a fraction of a square millimeter.

The source of radiation has a physical size. It casts its shadow (penumbral shadow), also called geometric unsharpness, behind the object. This shadow blurs the image. A radiographer cannot change the size of the radiation source, so this is a constant, beyond the control of the field personnel. However, either increasing the distance between the source and the specimen, or decreasing the space between film and specimen, or a combination of both, can reduce the effect of the penumbral shadow.

The changes in exposure conditions related to geometry – focus and projection – are interdependent in their effect on image sharpness. The object thickness also plays its role; objects that are too thick amplify the penumbral shadow. If the object under examination is positioned farther from the film, for a given source to film distance (SFD), the penumbral shadow will noticeably increase. The variables governing the penumbral shadow or 'geometric unsharpness' are mathematically related as follows:

$$U_g = FT/D$$

Where:

U_g = Geometric unsharpness
F = Focal spot size of the source
T = Specimen thickness
D = Source to specimen distance

The human eye can detect geometric unsharpness (U_g) up to 0.25 mm. Most good radiographs will have U_g of less than 1 mm (0.040 inch).

Film Graininess

Film graininess is the visual appearance of irregularly spaced grains of black metallic silver deposited in the finished radiograph. Radiographic films of all brands possess some degree of graininess. There are four major factors that determine the degree of graininess on a finished radiograph.

1. The type and speed of film.
2. The type of screen used.
3. The energy of radiation used.
4. The degree of development used.

Fluorescent intensifying screens produce what is termed 'screen mottle'. Screen mottle gives an appearance of graininess, considerably softer in outline than film graininess. This is a statistical variation in the absorption of radiation quanta by areas of the screen and the resulting fluctuations in intensification.

The energy level of the source usually affects film graininess. Higher energy radiation produces increased graininess. The effect has been attributed to electron scattering within the emulsion and subsequent sensitization adjacent to the silver halide grains.

Development or processing of the exposed radiograph beyond manufacturer's recommended time and temperatures can cause increased amounts of grain clumping, leading to a visual impression of film graininess.

IMAGE QUALITY INDICATOR (IQI) OR PENETRAMETER

As we have seen through this brief discussion of radiography, there are several variables that come together to give an acceptable level of image quality in the radiographic process. There is a need to develop a field-applicable quality assurance tool for consistent results, and for acceptable comparison of the quality of images produced by different technicians and through the use of different films.

Radiographers use a tool called a radiographic image quality indicator (IQI) or penetrameter. Figure 3-3-7 below shows a typical IQI. Penetrameters are manufactured in standard sizes of plate or wire in incremental thickness. The requirement varies from user to user, but the manufactured sets of IQIs cover most possibilities by overlapping the thickness of material being radiographed, and their composition covers the majority of materials used in construction. Materials are often grouped by their level of absorption, for example from Group I for lighter materials to Group V for heavy metals. The IQI material used must be selected from the same group as the specimen to be examined.

The image quality requirement is expressed in terms of penetrameter thickness, or for wire, the diameter of the visible wire. Most industrial radiography is required to have a sensitivity level of '2-2T'. The first 2 required that the penetrameter thickness (wire diameter, in the case of wire type IQI) is 2% of the specimen thickness. The second part (2T) together requires, on the plate type IQI, that a hole size that is twice the thickness of the IQI plaque be visible on the radiograph. The sensitivity of radiograph increases as the number is lowered.

RADIATION SAFETY

The effect of radiation on living cells is damaging, with the effect being dependent upon both the type and the energy of the radiation. For radiation safety the cumulative effect of radiation on the human body is of primary concern.

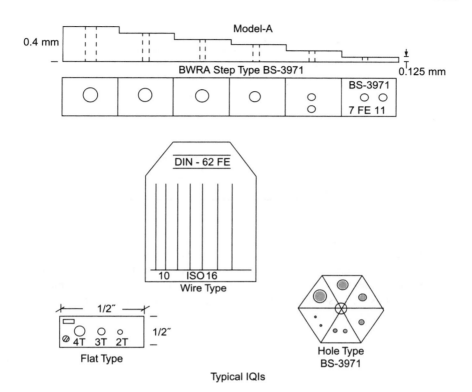

FIGURE 3-3-7 Typical (IQIs) penetrameter.

Exposure is measured in physical terms, and then a factor is used to compensate for the biological effects of different energies. We will now discuss briefly the various units used to measure radiation exposure.

Roentgen: The roentgen (r) is the unit that measures x-ray and gamma ray radiation in air. It is a physical measurement of x-ray and gamma ray (γ-ray) radiation quantity. It is defined as the quantity of radiation that would produce one electrostatic unit (esu) of charge in one cubic centimeter of air at standard pressure and temperature. One roentgen of radiation represents the absorption by ionization of approximately 83 ergs of radiation energy per gram of air. Practically, however, the milliroentgen (mr) is used, which is equal to 1/1,000 roentgen.

Rad: Radiation Absorbed Dose (Rad) is a unit of radiation absorbed by humans. It represents the absorption of 100 ergs of energy per gram of irradiated material at the place of exposure.

Rbe: Relative Biological Effectiveness (Rbe) is a value assigned to various types of radiation, determined by the radiation's effect on the human body. The National Committee on Radiation Protection has calculated Rbe values, which are given in the table below.

TABLE 3-3-7 RBE Values

Radiation	RBE values
X-Ray	1
Gamma Ray	1
Beta Particles	1
Thermal Neutrons	5
Fast Neutrons	10
Alpha Particles	20

The calculation of radiation safety level is easy as the unit used is the rem dose. For x-rays and gamma rays, it is a simple calculation, as the roentgen dose is equivalent to the rad dose, and the rbe of both x-rays and gamma rays are one. Thus a measurement of the roentgen dose becomes in effect a measurement of the rem dose.

Magnetic Particle Testing

Magnetic particle testing is a non-destructive method that detects discontinuities that are either buried slightly below or are open to the material or weld surface. Its advantage over visual inspection is that it can detect defects that are buried below the surface, and also surface opening defects that are too small to be visible to the naked eye. The sensitivity of the method is dependent on several factors, as described below.

The process is limited to such materials that have ferromagnetic properties, meaning that this method is not applicable to non-ferrous and non-magnetic material.

Since a wide variety of discontinuities can be located by this method, the training and experience of the tester is of great importance. For example, a person experienced in interpreting castings may not find it easy to interpret defects in welding and or wrought material; the same is true for an experienced weld defect interpreter if asked to interpret a casting.

The process requires some fundamentals to be correctly identified and agreed prior to the test. These are often written in referenced documents called specifications. They address the following essential points:

- Objective of testing
- Identification and description of test object
- Details of the testing techniques to be used

- Acceptance criteria
- Rework and retesting details.

This technique is not a substitute for radiography or ultrasonic testing. As stated at the beginning of this section, each NDT process has its advantages and limits. The engineer specifying the test must make a decision regarding the objective of testing and what types of discontinuity are expected. It is possible and practical to consider more than one NDT method to capture various types of discontinuities that may be expected in particular weldments.

PRINCIPLES OF MAGNETIC PARTICLE TESTING

When a magnetic field is established in a ferromagnetic material containing one or more discontinuities in the path of the magnetic flux, minute magnetic poles are set up at the opposite faces of the discontinuities. These poles attract the magnetic particles more strongly than the surrounding surface of the material, thus making it clearly visible against the contrasting background or under fluorescence of black light.

The most apparent characteristic of a magnet is its ability to attract any magnetic material placed within its field. This property is attributed to a line of force that passes through the magnetic materials, since they offer a path of lower reluctance than a path through the surrounding atmosphere, and these lines of force tend to converge in to the magnetic material.

The object for inspection is magnetized, either by an electromagnet or by a permanent magnet. In a permanent magnet system, the North and South Poles are at opposite ends; this is a longitudinal magnetization. These poles produce imaginary lines of force between them to create a magnetic field in the surrounding material.

This is explained by the following experiment with a bar magnet. If a bar magnet is notched as shown in Figure 3-4-1 below, the flux distribution or flow of the lines of force will be markedly changed in the area surrounding the notch. The distortion in the line diminishes as the distance from the breach in the magnetic field increases. In this condition each face of the notch assumes an opposite polarity, producing a flow of leakage flux across the air gap. It is this leakage flux that permits the detection of defects by the magnetic particle method. Irrespective of what method is used to magnetize the test object, its magnetic field remains the principle of the method.

The electromagnetic method is used more often. In this method the test object is magnetized by introducing a high current, or by putting the test object in a current-carrying coil. The magnetic field in the test piece is interrupted by any discontinuities, producing a magnetic field leakage on the surface. The area to be inspected is covered by finely divided magnetic particles that react to magnetic field leakage produced by the discontinuity. The magnetic particles form a pattern of indication on the surface, as the magnetization of these particles assumes the approximate shape of the discontinuity.

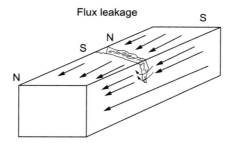

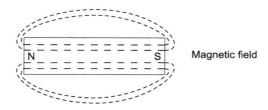

FIGURE 3-4-1   Magnetic field and flux leakage.

Some of the commonly used methods for magnetizing a test object are based on how the magnetizing field in a given object will be induced and managed to the best advantage. We shall now briefly describe them:

- Longitudinal magnetizing using coils
- Circular magnetizing using coils
- Magnetizing by use of current carrying conductor
- Prod magnetization.

CALCULATING MAGNETIZING CURRENT

The importance of the correct amount of current to be used for testing cannot be over-emphasized. Just to illustrate the point, two extreme conditions are listed below. A current that is too small will produce a low magnetic field which will not be able to detect all possible discontinuities; however, too large a current may produce too strong magnetic fields that would mask the small and tightly adhering faces of a crack.

The magnetic current should be established by standards and specifications, or given within the purchase order. In the situation where a specification is not available, the current requirement may be calculated as given below:

1. Longitudinal magnetizing: 3,000 to 10,000 amperes turn.
2. Overall circular magnetizing: 100 to 1,000 amperes per inch (or 4 to 40 A/mm) of the outside diameter of the part.

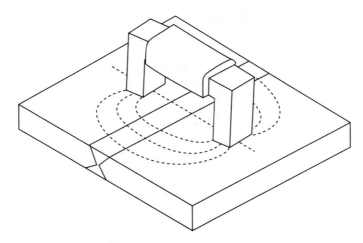

FIGURE 3-4-2   Longitudinal magnetic field developed by electromagnet.

3. Prod magnetization: 100 to 125 amperes per inch (or 4 to 5 A/mm) of the prod spacing.

- Dry magnetization: Prod spacing of 3 to 12 inch (75 mm to 340 mm)
- Wet method: Prod spacing of 2 to 12 inch (50 mm to 340 mm)
- Typical longitudinal magnetizing with electromagnetic yoke is shown below.

TYPES OF MAGNETIZING CURRENT

High ampere, low voltage current is commonly used for testing. It can be AC, DC or rectified current. Most portable equipment used for field-testing either uses permanent magnets or electromagnets, but these have limitations, as they are capable of detecting surface opening defects they are not very efficient at detecting subsurface defects.

Alternating current (AC) is used where subsurface evaluation is not required, as only the surface of the test material is magnetized. This method is effective for locating surface opening discontinuities.

Direct current (DC) magnetization produces a field that penetrates throughout the part, and as a result it is more capable of detecting subsurface discontinuities. Three phase, full wave, rectified current produces results comparable to those obtained by battery-powered magnetization.

Half-wave rectified single phase current provides maximum sensitivity. The pulsating field increases particle mobility and enables the particles to line up more readily in weak leakage fields. The pulsed peaks also produce a higher magnetization force.

INSPECTION METHOD

The specified sequence of operations for a successful test process is listed below:

1. Pre-cleaning of the test surface.
2. Application of the current continuous or residual application.
3. Application of inspection media, dry, wet, color contrast, florescent etc.
4. Inspection method and evaluation, use of visual aids, including ultraviolet (black) light.
5. Demagnetizing residual magnetism.
6. Reporting.

Pre-Cleaning of Test Surface

Surface preparation is very important for obtaining satisfactory results, since the condition of the surface is important. The material to be tested may be in as-welded, as-rolled, as-cast, or as-forged condition, and hence the test surface may contain various process contaminants that may mask the test result or mislead its interpretation. The surfaces of the parts to be tested are therefore prepared by grinding, machining, solvent or water cleaning, or other methods that do not block the surface opening of any possible discontinuities.

Prior to the magnetic particle examination, an area about 100 mm around the actual test area on the surface is visually examined to ensure that it is dry, and free of all dirt, grease, lint, scale, welding flux, weld spatter, paint, oil, and other extraneous matter.

Drying after Preparation

The surfaces to be examined are dried after cleaning. Drying is done by normal evaporation or by blowing with forced hot or cold ambient ($\approx 20°C$) air. It must be ensured that the cleaning solution has evaporated prior to application of the magnetizing current and magnetic particles.

Application of the Current

High amperage, low voltage current is usually used, but alternating current (AC), direct current (DC) or rectified current can be used for magnetizing the parts to be inspected.

Portable equipment that makes use of an electromagnet or permanent magnets are also used. Both of these methods can only detect surface cracks.

Alternating Current

When an alternator produces an alternating current voltage, the voltage switches polarity over time, in a very particular manner. If this polarity over time wave trace is graphed, it is seen to change rapidly but smoothly over the cross over line (point zero). The shape of the curve so produced is called a sine wave.

One cycle of this reversal is termed a wave cycle, and the rate of this alternation is called the frequency, which is measured in Hertz (Hz). In the USA, the electrical supply from the grid is at 60 Hz, while in most of Europe the frequency is 50 Hz.

Due to the sine wave pattern of current, the current can only penetrate the surface, thus only the surface of the metal is magnetized by alternating current. This method is effective for locating discontinuities that extend to the surface, such as fatigue or service cracks. Similarly, in a weld, a surface opening can be detected by AC current magnetization. As stated in the introduction, this type of magnetizing is not able to detect subsurface discontinuities.

Direct Current

Unlike AC current, the wave of travel of direct current is linear and flat. This means that the magnetization is able to penetrate through the material, hence it is more sensitive for the detection of subsurface discontinuities.

Full-wave, three-phase, rectified current produce results essentially comparable to the direct current obtained from batteries.

Half-wave, single-phase rectified current provides maximum sensitivity. The pulsating field increases particle mobility, and enables the particles to line up more readily in weak leakage fields. The pulse peaks also produce a higher magnetizing force.

The operational sequence of magnetization and application of the inspection medium has an important bearing on the sensitivity of the method.

Continuous or Residual Application of Current

The continuous method is the process in which the magnetizing and application of dry-magnetic powder is simultaneous. In fact, the application of medium (dry-powder) is done only as the current is flowing through the test material.

This condition offers maximum sensitivity, as the magnetic field is at its peak during the entire process. If the current is stopped during the application of the medium, only the residual current will be able to hold the indications.

Where a wet suspension is used with the continuous method, the medium in wet suspension is allowed to flow over the material as the magnetizing current is applied. The inspection is carried out as the medium is applied, and current is applied in bursts of about one-half second.

The residual method, as its name implies, depends on the residual magnetic field to detect discontinuities. The current is applied, and magnetic fields develop, and then the current is switched off as the indicating medium is applied. Accuracy and sensitivity depend on the strength of the residual magnetic field. It is obvious that the application of this method is most effective on those materials that have higher magnetic retentivity.

Knowledge of the magnetic permeability and retentivity of the material being magnetized and or tested helps to establish the best method for magnetizing and testing. If available, the hysteresis curve of the particular metal must be reviewed. This will help develop an effective magnetizing or demagnetizing procedure.

A hysteresis curve is a plot of flux density versus magnetic force. A hysteresis curve can be either a wide or a slender looped curve; each signifies specific properties of the magnet. The following are typical characteristics of a hysteresis curve.

A wide loop curve signifies:

- Low permeability of the metal being magnetized
- High magnetic retentivity of magnetized material
- High coercive force, required to remove magnetism from the material
- High reluctance (difficult to magnetize)
- High residual magnetism.

A slender loop curve signifies:

- High permeability of the metal being magnetized
- Low magnetic retentivity of magnetizing material
- Low coercive force, required to remove magnetism from the material
- Low reluctance (easy to magnetize)
- Low residual magnetism.

DRY METHOD OF INSPECTION

As the name suggests, the method is based on the use of dry powder media for inspection. The method uses finely divided ferromagnetic particles in dry powder form. The particles are coated for greater mobility, and they are uniformly dusted over the inspection surface, using dusting bags, atomizers or spray guns. These particles have a variety of colors to easily distinguish the patterns formed from the background while inspecting.

WET METHOD OF INSPECTION

In the wet magnetic particle inspection method, the material to be inspected is sprayed, flowed or immersed in a bath of the suspension. The method uses both color contrast and fluorescence under ultraviolet (black) light.

The size of particles used for the wet method is smaller than with dry powder. This is because the particles are suspended in a liquid bath of light petroleum distillate. These particles are supplied either in dry concentrate form or in a paste form, and the final formulation is prepared with either an oil or water bath. The specific formulations generally cannot be changed from oil to water base. The smaller size of particles makes the wet method more sensitive to fine surface defects, but is limited in detecting subsurface defects. When the bath method is used, it needs to be agitated continuously to prevent the particles settling. The water-based suspension is preferable, as it is as sensitive as the oil-based version, but is not a fire hazard. Flammable liquids like kerosene oil are often the base for a magnetic particle bath.

VIEWING CONDITIONS

Non-fluorescent wet particle testing requires that the parts being inspected are illuminated to at least 200 foot candles (2,152 Lux) of visible light.

When using fluorescent material, it is necessary that ultraviolet light is used to view the indications presented by fluorescent particles.

Inspection under Ultraviolet (Black) Light

Ultraviolet light supplies the correct wavelengths to cause fluorescent material to fluoresce. The equipment essentially consists of a regulating transformer, a mercury arc lamp, and a filter. The mercury arc bulb and the filter are contained in the reflector lamp housing and the transformer is housed separately. A deep red-purple filter is designed to transmit only those wavelengths of light that will activate the fluorescent material.

For correct test results, the lamp should be able to produce a minimum intensity of $800\,\mu w/cm^2$ (microwatts per centimeter squared), in a three-inch circle at a 15-inch distance.

Once the switch is turned on, it takes about five minutes for the light to attain its full intensity. Once turned on the light is kept 'ON' for the entire duration of the test, to keep the light ready for inspection without interruptions, and also because frequent switching on and off shortens the life of the arc bulb.

The dust and dirt sticking on the filter glass significantly reduces the intensity of the light. This means that the filter glass should always be kept clean.

Penetrant Testing

The basic principle of liquid penetrant testing (PT) is capillary action, which allows the penetrant to enter in the opening of the defect, remain there when the liquid is removed from the material surface, and then re-emerge on the surface on application of a developer, which has a capillary action similar to blotting paper. If done properly, it is a very effective test method that exposes surface-opening discontinuities for visual inspection. Inspection can be carried out either without any additional visual aid, or if a fluorescent penetrant is used, under ultraviolet black light. The process is excellent for non-magnetic materials. Very small and tight imperfections can easily be detected.

Applied Welding Engineering: Processes, Codes and Standards.

The method has several variants, divided into two basic categories:

(a) Fluorescent and
(b) Visible dye methods.

Each has their advantages; we shall discuss them further in this section.

GENERAL PROCEDURE

A liquid penetrant examination is performed in accordance with a written procedure. This procedure is based on the specific requirements of the job, and considers at least the following details.

- The objective of the examination.
- The shapes and size of the material to be examined, and the extent of the examination.
- The type of each penetrant, penetrant remover, emulsifier, and developer to be used; these are specifically identified by product-specific number or letter designators.
- Details of processing of each stage are included. For example, for pre-examination cleaning and drying, the dwell time, application method and temperature range, cleaning materials to be used and minimum time allowed for drying are all specified.
- The processing details for removing excess penetrant from the surface, and for drying the surface before applying the developer.
- The processing details for applying the developer, and length of developing time before interpretation.
- The processing details for post-examination cleaning.

Any change in part processing that can close surface openings of discontinuities or leave interfering deposits, such as the use of grit blast cleaning or acid treatments must be identified and remedial action must be suggested.

PENETRANT MATERIALS

In general use, the term penetrant material includes all penetrants, solvents or cleaning agents that are used in this examination process. A penetrant material has the capacity to enter the crevices opening on the surface of a material. The material must have good wetting properties, and the penetrant must be suitable for capillary action. If a developer is used, the penetrant material must be able to be absorbed by the developer in uniform manner.

SPECIFIC REQUIREMENTS

Control of Contaminants

The certification of contaminant content for all liquid penetrant materials used on nickel-based alloys, austenitic stainless steels, and titanium shall be obtained and reviewed. These certifications must include the penetrant manu-facturer's batch numbers and the test results obtained in accordance with (a) and (b) below.

These records shall be maintained for review as required.

(a) When examining nickel base alloys, all materials shall be analyzed indi-vidually for sulfur content as follows:

(1) An individual sample of the penetrant materials, with the exception of cleaners, shall be prepared for analysis by heating 50 g of the material in a 150 mm nominal diameter glass petri dish at a temperature of 194°F to 212°F (90°C to 100°C) for 60 min. This must be done in a properly ventilated laboratory to dissipate the emitted vapor generated by the process. The residue is evaluated and analyzed to meet following limits to be acceptable.
   • The residue should be ≤0.0025 g.
   • The sulfur content should be ≤1% of the residue by weight.

(2) The cleaner and remover material is similarly tested. A sample of mate-rial is prepared for analysis by heating 100 g of the material in a 150 mm nominal diameter glass petri dish at a temperature of 194°F to 212°F (90°C to 100°C) for 60 min. The residue is evaluated and analyzed to meet the following limits.
   • The residue should be ≤0.005 g for the material to be acceptable.
   • The sulfur content should be ≤1% of the residue by weight.

(b) For examining austenitic stainless steel or titanium, the importance of con-taminants like chlorine and fluorine become important. The following tests are carried out to determine these contaminants.

(1) The sample of penetrant material is prepared for analysis. A sample of 50 g in a 150 mm diameter glass petri dish is heated to a temperature of 194°F to 212°F (90°C to 100°C) for 60 min. If the residue is ≤0.0025 g, the penetrant is acceptable. ASME SE 165 gives alternative test methods for testing for chlorine and fluorine where the combined residue will be less than 1% by weight.

(2) Similarly the 100 g sample of cleaner material in 150 mm glass petri dish is tested at a temperature of 194°F to 212°F (90°C to 100°C) for 60 min. The acceptable amount of residue is ≤0.005 g. ASME SE165 gives another alternative technique using ion chromatography for pen-etrant testing cleaner/remover material for chloride and fluoride.

Surface Preparation

Surface preparation is very important for obtaining satisfactory penetrant test-ing results.

(a) For parts to be tested in the as-welded, as-rolled, as-cast, or as-forged con-dition, the surface is prepared by grinding, machining, or other methods that are not capable of blocking the surface openings of the possible discontinuities.

(b) Prior to the liquid penetrant examination, the target area and about 25 mm of the adjacent area on the surface is visually examined to ensure that the material is dry and free of all dirt, grease, lint, scale, welding flux, weld spatter, paint, oil, or other extraneous matter that could obscure surface openings or otherwise interfere with the examination.

(c) The cleaning agents used to remove contaminants from the surface are detergents, organic solvents, descaling solutions, paint removers etc., as the requirement of the specific part to be examined; more than one agent may be used in some cases. Degreasing and ultrasonic cleaning methods are also used.

Drying after Preparation

Surfaces need to be examined are dried after cleaning. Drying is done by normal evaporation or with forced hot or cold air. It must be ensured that the cleaning solution has evaporated prior to application of the penetrant.

TECHNIQUES

Fluorescent or visible penetrant with color contrast are used with one of the following three penetrant processes:

(a) Water washable
(b) Post-emulsifying
(c) Solvent removable

The combination of fluorescent or visible penetrant with the three processes listed above results in six possible liquid penetrant techniques.

Techniques for Standard Temperatures

The technique is generally applied to material surfaces between 10°C (50°F) and 50°C (125°F). This temperature range is maintained throughout the testing period.

Penetrant Application

The penetrant is applied by a suitable means, such as:

(a) Dipping the component in the penetrant,
(b) Brushing the penetrant on the surface to be examined or
(c) Spraying the penetrant on the surface to be examined.

If spraying is used, proper filters must be used to control contaminants like oil, water, dirt, or sediment that may have collected in the compressed air lines.

Penetration Time (Dwell Time)

Penetration time is critical. The minimum penetration time is often established and is generally regulated by the testing code. General guidelines on dwell times are given in Table 3-5-1.

Excess Penetrant Removal

After the specified penetration time has elapsed, any remaining penetrant on the material surface is removed. Care is taken to ensure that only surface excess is removed and no attempt is made to remove penetrant from disconti-nuities. Different penetrants require specific care in their removal.

Removing Excess Water-Washable Penetrant

As the name implies, the excess water-washable penetrant is removed by spraying with water. The water pressure is kept below 50 psi (345 kPa), and the water temperature is maintained below 110°F (43°C). Care must be taken to remove only the excess penetrant on the surface and not from the crevices and cracks that need to be evaluated.

Removing Excess Post-Emulsifying Penetrant

The post-emulsifying penetrant is applied by spraying or dipping. The emulsi-fication time is critical. The dwell time is governed by surface roughness and type of applied emulsifier. The dwell time is qualified by actual tests. After emulsification, the mixture is removed by a water spray, similar to the proce-dure used for water-washable penetrant.

Removing Excess Solvent-Removable Penetrant

Excess solvent-removable penetrant is removed by wiping with a lint-free cloth or absorbent paper, and repeating the operation until most traces of pen-etrant are removed from the surface. The remaining traces are removed by lightly wiping the surface with lint-free cloth or absorbent paper moistened with solvent. Excess solvent must be avoided to minimize the removal of pen-etrant from discontinuities.

Drying Process after Excess Penetrant Removal

In the water-washable and post-emulsifying penetrant process, the surface can be dried by blotting with clean materials or by using circulating air, keeping the temperature of the surface below 125°F (50°C).

Drying in the solvent-removable process is done by forced air or normal evaporation, blotting, or by wiping the surface with lint-free cloth or paper.

Developing

The developer is applied immediately following the removal of the excess penetrant from the surface. A uniform and reasonable layer of the developer is applied. Insufficient coating thickness may not draw the penetrant out of discontinuities; conversely, excessive coating thickness may mask indications.

When using a color contrast penetrant, only a wet developer is used. If however a fluorescent penetrant is selected then either wet or dry developer is used.

Dry Developer Application

Dry developer is applied on a dry surface by a soft brush, hand powder bulb, powder gun, or other means. The object is to dust the powder evenly over the entire surface to be examined.

Wet Developer Application

The suspension-type wet developer must be properly shaken prior to application. This ensures adequate dispersion of suspended particles.

Application of Aqueous Developer

Aqueous developer is applied to either a wet or dry surface. The developer is applied either by dipping, brushing, spraying, or other means, provided that a thin layer of coating is obtained over the entire surface to be examined.

Application of Non-Aqueous Developer

Non-aqueous developer is applied only to a dry surface, usually by spraying. If, due to safety reasons, spraying is not possible then the developer is applied by brushing. Normal evaporation processes are used to dry the developer.

The developing time for final interpretation starts immediately after the application of a dry developer or as soon as a wet developer coating is dry. The minimum developing time is given in Table 3-5-1.

INTERPRETATION

Final Interpretation

Final interpretation needs to be made within 7 to 60 minutes after the dwell time of developer. If bleed-out does not alter the examination results, longer periods are permitted. If the surface to be examined is too large to allow complete examination within the prescribed or established time, the examination is performed in increments.

Characterizing Indication(s)

Some types of discontinuities are difficult to evaluate, especially if the penetrant diffuses excessively into the developer. In such conditions, the application and interpretation must be carried out simultaneously to observe the

TABLE 3-5-1 Minimum Dwell Time

Material	Form of material	Type of targeted Discontinuity	Dwell Time (in minutes) (at 10°C to 50°C)	
			Peneterant	Developer
Aluminum, Magnesium Steel Brass Bronze, Titanium High temperature alloys	Casting and Welds	Cold shuts, Porosities, lack of fusion, all types of cracks	5	7
	Wrought material-Plates, Forgings Extrusions	Laps, all forms of cracks	10	7
Carbide tipped tools		Lack of fusion, Porosities, cracks	5	7
Plastics	All forms	Cracks	5	7
Glass	All forms	Cracks	5	7
Ceramics	All forms	Cracks, Porosity	5	7

formation of indication(s). This is helpful in characterizing the nature of the indication and its extent.

Color Contrast Penetrant

In the color contrast penetrant process, the developer forms a reasonably uniform white coating. Surface discontinuities are indicated by bleed-out of the penetrant, which is normally a deep red color that stains the developer. Indications with a light pink color may indicate excessive cleaning, which may leave an excessive background making interpretation difficult. A minimum light intensity of 50 fc (500 Lx) is required to ensure adequate sensitivity during the examination and evaluation of indications.

Fluorescent Penetrant

The fluorescent penetrant process is similar to the color contrast process except that the examination is performed using ultraviolet light, which is also called black light.

The examination process is sequenced below.

(a) The test is performed in a dark room or dark enclosure.

(b) The examiner is in the dark area for at least 1 minute prior to performing the examination to enable his eyes to adapt to dark viewing. If the examiner wears glasses or lenses, they should not be photosensitive.

(c) The black light is allowed to warm up for a minimum of 5 minutes prior to use or measurement of the intensity of the ultraviolet light emitted.

(d) The black light intensity is measured with a black light meter. A minimum light intensity of $1000\,mW/cm^2$ on the surface of the part being examined is required. The black light intensity is measured at least once every 8 hours, and whenever the workstation is changed.

EVALUATION

All indications are evaluated in terms of the acceptance standards of the applicable inspection code.

The interpreter needs to be able to distinguish the discontinuities at the surface that is indicated by bleed-out of penetrant; and also the false indications caused by the localized surface irregularities due to machining marks or other surface conditions.

Broad areas of fluorescence or pigmentation should be avoided, as they tend to mask indications of discontinuities. If such conditions exist, the affected area should be cleaned out and re-examined.

Fluorescent penetrant examination should not follow a color contrast penetrant examination. Intermixing of penetrant materials from different families or different manufacturers should not be done.

Liquid Penetrant Comparator

Testing Beyond the 10°C to 52°C Temperature Range

Liquid penetrant comparator blocks are used when the test needs to be conducted outside the 50°F to 125°F (10°C to 52°C) temperature range.

The process involves comparator blocks made of ⅜ inch thick rectangular aluminum plate of specified grade.

The Procedure

On the selected test block, an area is marked at the center of both faces. The marked area is approximately 1 inch (25 mm) in diameter. The mark is made with a 950°F (510°C) temperature-indicating crayon. It is then heated with a torch or burner to a temperature between 950°F (510°C) and 975°F (524°C). The specimen is then immediately quenched in cold water. This produces a network of fine cracks on each face. After cooling, the block is cut in half. One-half of the specimen is marked as block 'A' and the other as block 'B', for identification.

Comparator Application

If a liquid penetrant examination procedure is to be carried out at a temperature of less than 50°F (10°C), the proposed procedure should be applied to block 'B' after the block and all materials have been cooled and held at the proposed examination temperature until the comparison is completed.

The standard procedure that was previously used to demonstrate at block 'A' in the 50°F to 125°F (10°C to 52°C) temperature range on block 'A' is taken as reference for comparison.

Indications of cracks from both samples A and B are then compared. If the indications obtained under the proposed conditions on block 'B' are essentially the same as those obtained on block 'A' during examination at 50°F to 125°F (10°C to 52°C), then the proposed procedure should be considered to be qualified for use.

Ultrasonic Testing

This chapter provides you with an introduction to the world of sound.

Physicists and acoustic engineers tend to discuss sound pressure levels in terms of frequencies, partly because this is how our ears interpret sound. What we experience as 'higher pitched' or 'lower pitched' sounds are pressure vibrations having a higher or lower number of cycles per second. In acoustic measurement, acoustic signals are sampled over time, and then presented in more meaningful forms such as octave bands or time frequency plots. Both these popular methods are used to analyze sound and better understand acoustic phenomena.

The fundamental principles being same, there are more advanced developments in the field and basic principles are dissected and utilized to address more specific needs of this ever-growing industry. Some of these specialized 'subsections' or branches or new developments in ultrasonic testing are listed below. Some of these will be addressed in this chapter, but for details readers are directed to specialized suppliers, as some are very proprietary techniques:

- UT – Shear Wave Ultrasonic Testing
- AUT – Automated Ultrasonic Testing
- PAUT – Phased Array Ultrasonic Testing
- TOFD – Time of Flight Diffraction Ultrasonic Testing
- GWUT – Guided Wave Ultrasonic Testing

- LGWUT – Long Range Guided Wave Testing
- SWUT – Surface Wave Ultrasonic Testing technology.

THEORY OF SOUND WAVE AND PROPAGATION

In fluids such as air and water, sound waves propagate as disturbances in the ambient pressure level. While this disturbance is usually small, it is still audible to the human ear. The smallest sound that a person can hear, known as the threshold of hearing, is nine orders of magnitude smaller than the ambient pressure. The loudness of these disturbances is called the sound pressure level, and it is measured on a logarithmic scale in decibels. Some common examples of sound and pressure amplitudes and their decibel levels are given in Table 3-6-1. Mathematically, the sound pressure level (SPL) is defined as:

$$SPL = 20 \times \log_{10}(P/P_{ref})$$

Where:

P_{ref} is the threshold of hearing and
P is the change in pressure from the ambient pressure.

The following table gives a few examples of sounds and their strengths in decibels and Pascals.

The entire sound spectrum can be divided into three sections: audio, ultrasonic, and infrasonic. The audio range falls between 20 Hz and 20,000 Hz. This range is important because these frequencies can be detected by the human ear. This range has a number of applications, including speech, communication and music.

The ultrasonic range refers to the very high frequencies: 20,000 Hz and higher. This range has shorter wavelengths which allow better resolution in imaging technologies. Industrial and medical applications such as ultrasonography and elastography rely on the ultrasonic frequency range.

TABLE 3-6-1 Pressure Amplitude and Decibel Level

Example of Common Sound	Pressure Amplitude	Decibel Level
Threshold of hearing	$20*10^{-6}$ Pa	0 dB
Normal talking at 1 m	0.002 to 0.02 Pa	40 to 60 dB
Power lawnmower at 1 m	2 Pa	100 dB
Threshold of pain	200 Pa	134 dB

On the other end of the spectrum, the lowest frequencies are known as the infrasonic range. These frequencies can be used to study geological phenomena such as earthquakes etc.

As stated in the introduction above, the term ultrasonic is the name given to the study and application of sound waves having frequencies higher than those that can be heard by human ears.

Ultrasonic non-destructive testing is the use of the ultrasonic sound spectrum to examine or test materials, or to measure thickness, without destroying the material. The testing frequencies range from 100,000 cycles per second (100 kHz) to 25,000,000 cycles per second (25 MHz).

Ultrasonic testing does not give direct information about the exact nature of the discontinuity. This is deduced from a variety of information, like material properties and construction.

THEORY OF SOUND

Sound is the mechanical vibration of particles in a medium. When a sound wave is introduced into a material, the particles in the material vibrate about a fixed point at the same frequency as the sound wave. The particles do not travel with the wave but react to the energy of the wave. It is the energy of the wave that moves through the material.

The length of a particular sound wave is measured from trough to trough, or from crest to crest. This distance is always the same. This distance is known as the wavelength (λ). The time taken for the wave to travel a distance of one complete wavelength (λ) is the same amount of time it takes for the source to execute one complete vibration. The velocity of sound (V) is given by the following equation:

$$V = \lambda * F$$

Where:

λ = The wavelength of the wave
F = The frequency of the wave.

A number of sound waves travel through solid matter. Some of them are listed below.

Longitudinal waves, also called compression waves. The particles vibrate back and forth in the same direction as the motion of the sound. The ultrasonic vibrations in liquids and gases only propagate in longitudinal waves. This is because liquids and gases have no shear rigidity.

Shear waves, also called transverse waves. The particles in this type of wave vibrate back and forth in a direction that is at right angles to the motion of the sound.

It is also possible in some specific circumstances, to produce shear waves that travel along the free boundary or surface of a solid. These **surface** or **Rayleigh waves** penetrate the material to a depth of only a few particles.

In solids all the three modes of sound waves can propagate.

The shortest ultrasonic wavelengths are of the order of magnitude of the wavelength of visible light. Because of this, ultrasonic wave vibrations possess properties which are very similar to light waves; that is, they can be reflected, focused, and refracted.

High frequency particle vibrations of sound waves are propagated in homogeneous solids in the same manner as directed light beams. Sound beams are reflected either partially or totally at any surface acting as a boundary between the object and the gas, liquid or other type of solid. Ultrasonic pulses reflect from discontinuities, thereby enabling detection of their presence and location. Some of the terminology particularly associated with ultrasonic testing methods is described below.

PIEZOELECTRICITY

Piezoelectricity refers to the electricity produced by a vibrating crystal and its reversion back to vibrations of the crystal.

When an electric current is applied to the crystal, it transforms the electrical energy into mechanical vibrations and transmits them through a coupling medium into the test material. These pulsed vibrations propagate through the object with a velocity that depends on the density and elasticity of material.

SOUND BEAM REFLECTION

High frequency sound waves act in a similar way to light waves. If the wave is interrupted by an object, most of the sound beam is reflected. Crystals or transducers then pick up these sound beams and present them as vertical deflections of a horizontal trace or a line base on a cathode ray tube (CRT) or an oscilloscope. This type of presentation is called A-scan, other presentations being the B-scan, which presents a cross-sectional image of the discontinuity and the material being inspected. The C-scan presentation displays the discontinuity in plan-view.

SOUND BEAM FREQUENCIES

Most ultrasonic testing works between 400 kHz and 25 MHz. These vibrations are beyond the audible range and propagate in the test material as waves of particle vibrations. Sound beams of all frequencies can penetrate fine-grained material without difficulty. When using high frequencies in coarse-grained material, interpretation becomes difficult as interference in the form of scattering is noted. Depth of penetration is better achieved by lower frequencies.

The selection of a specific frequency for testing is mainly dependent on material properties and the goal of the test.

Frequencies of up to 1 MHz are generally a good choice as they have better penetration, have less attenuation and they are scattered less by coarse grains and rough surfaces. The disadvantage of low frequencies is that they have a large angle of divergence, and as such they can't resolve small flaws.

On the other hand, the high frequency transducers emit more concentrated beams with better resolving power, but they are scattered more by coarse grains and rough surfaces.

Frequencies above 10 MHz are normally not used in contact testing, because the higher frequency transducers are thinner and more fragile. As the frequency of sound vibrations increases, the wavelength correspondingly decreases and approaches the dimensions of a molecular or atomic structure. In immersion testing, however, all frequencies can be used because there is no physical contact between the transducer and the material being tested.

SOUND BEAM VELOCITIES

Ultrasonic waves travel through solids and liquids at relatively high speeds, but they are relatively rapidly attenuated or die down. The velocity of a specific mode of sound is a constant through a given homogeneous material. Ultrasonic wave velocities through various materials are given in Table 3-6-2 below. The difference in sound velocity is due to the density and elasticity of each material. Yet, density alone is not able to account for the whole of the variation, as it may be noted that beryllium has high sound velocity, although it is less dense than aluminum, and also the acoustic velocities of water and mercury are nearly the same, although the density of mercury is 13 times greater than that of water.

TABLE 3-6-2 Sound Velocities in Various Mediums

Material	Density g/cm^3	Longitudinal Velocity	
		cm/µSec	
Air	0.001	0.033	738
Water	1.00	0.149	3,333
Plastic (Acrylic)	1.18	0.267	5,972
Aluminum	2.8	0.625	13,981
Steel			
Cast Iron			
Mercury	13.00	0.142	3,176
Beryllium	1.82	1.28	28,633

While discussing the properties of sound we noted that sound beams are refracted and subjected to mode conversion, resulting in a combination of shear and longitudinal waves. The question arises as to what are the principles governing the mode of transformation.

When a longitudinal ultrasonic wave is directed from one medium to another of different acoustic properties, which is arranged at an angle other than normal to the interface between the two media, a wave mode transformation occurs. The resultant transformation depends on the incident angle in the first medium and on the velocities of sound in the first and second media. In each transformation, there is an equal angle of reflection back into the first medium. Snell's law is used to calculate angle transformations based on the sound path angles and the sound velocities of the two media.

SNELL'S LAW OF REFLECTION AND REFRACTION

Figure 3-6-1 shows an illustration of Snell's law of reflection and refraction. We note that the sine of the incident angle a is to the sine of d (longitudinal) or e (shear) refracted angle as sound velocity of the incident medium 1 is to the sound velocity of the refracted medium 2. The same relationship exists when mode conversion occurs within the same medium, using the sound velocities of different waves in the equation. This can be used to calculate refracted and reflected angles. For part (a) of the sketch, these equations can be written as:

$$\sin a / \sin d = \text{Longitudinal velocity in medium-1} / \text{Longitudinal velocity in medium-2}$$

$$\sin a / \sin e = \text{Longitudinal velocity in medium-1} / \text{Shear velocity in medium-2}$$

or

$$\sin a \,(\text{long}) / \sin c \,(\text{shear}) = \text{Longitudinal velocity in medium-1} / \text{Shear velocity in medium-1}$$

In part (b) of the sketch, the equation will be:

$$\sin a \,(\text{shear}) / \sin c \,(\text{long}) = \text{Shear velocity in medium-2} / \text{Longitudinal velocity in medium-2}$$

We note that as the incident angle is increased from the normal, this results in only a longitudinal wave. This longitudinal angle d also increases until it reaches 90°. At that point no more longitudinal wave enters the second medium. This angle of the medium is called **First Critical Angle**.

As the incident angle is further increased, the shear angle e also increases until it becomes 90°. At this point the entire shear wave in the second medium is transformed into the surface wave. This is called **Second Critical Angle**.

These calculations use a simple centerline of the beam as input, but the actual application is more complex, however, as the sound beam has width and

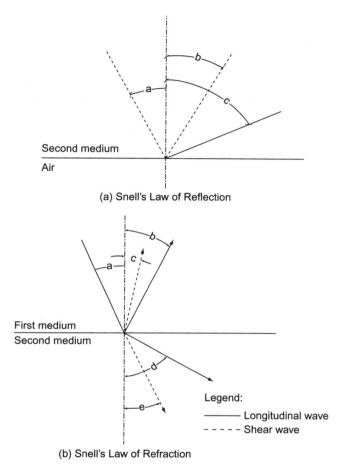

(a) Snell's Law of Reflection

(b) Snell's Law of Refraction

Legend:
———— Longitudinal wave
- - - - Shear wave

FIGURE 3-6-1   Snell's Law of Reflection and Refraction.

divergence. The amplitude of the sound is also higher at its centerline and it gradually dies down on the outer edges.

Understanding the Variables Associated with Ultrasonic Testing

The sound velocity through a given material is the distance that the sound energy will propagate in that material in a given time, and it is a function of material density, acoustic impedance and temperature.

Since sound velocities are relatively high, most values are given in meters or feet per second. The sound velocity of a shear wave in a given material is usually one half of a longitudinal and about 1.1 times that of a surface wave. Specific velocities are tabulated in handbooks and are used for calculations to determine the angle transformations.

The effect of temperature on sound velocity is normally not very significant in most metals but must be considered when calculating angles in plastics if they are used as a wedge for shear wave research units.

The frequency of ultrasound used for testing is usually between 1 and 6 MHz. The most common frequency for weld inspection is 2.25 MHz. The transducer element when exited resonates at its natural frequency. The resulting frequency is not a single frequency but a relatively narrow band of frequencies. One or more of these will respond with highest amplitude. The frequency is related to the thickness of the transducer element. Frequency decreases as the thickness of the element is increased.

The piezoelectric property of the transducer is affected by the natural frequency of the element, and must be considered to obtain maximum sound amplitude. It may be noted that broadband pulse generation is usually effective and used with portable equipment.

As explained earlier in the theory of sound, the wavelength (λ) is the function of velocity (V) and frequency (F):

$$V = \lambda * F$$

The expected minimum size of reflector (flaw) that is detectable with ultrasonic sound is about one half wave length ($\frac{1}{2}$) as measured in a direction perpendicular to the direction of sound propagation.

SELECTION OF TEST EQUIPMENT

A variety of ultrasonic test applications have prompted industry to develop specialized test equipment. The enormous development of electronics has allowed manufacturers to add more sophisticated functional features to their equipment, increasing its portability, improving its results and taking away several calculations from the hands of the operators and adding them to the machine as features. Miniaturized size and higher speed are very common features of new equipment. Other developments include:

- Typically, real time 320 × 240 pixel (QVGA)
- A-trace (40 Hz update) display
- Distance amplitude flaw gating (DAG)
- Weld trig
- Freeze and insta-freeze for spot weld applications
- Precision thickness measurement capability, combined flaw & thickness mode
- 2 Megabyte of memory for substantial storage capability, USB connectivity
- High speed scrolling & encoded B-scan
- IP & IF gating
- Adjustable damping, multi-color LCD screens
- Other features such as SplitView, SplitScan, AutoTrack and Quarter VGA resolution etc. are also offered by manufacturers.

However, the basic ultrasonic principles remain the same, and the available equipment is based on these principles. Longitudinal wave ultrasound is generally limited in use to detecting inclusions and lamellar type discontinuities in the base material. Shear waves are the most valuable in detecting weld discontinuities because of their ability to give their three-dimensional coordinates. As stated earlier in this section, the sensitivity of a shear wave is about twice that of a longitudinal wave, the frequency and search unit size being constant.

Shear wave angles are measured in the test material from a line perpendicular to the test surface. Selection of the search unit's angle is based on the expected flaw orientation. Usually it is a good practice to use more than one search angle to ensure proper flaw detection. The three most common angles used for shear wave testing are 70°, 60°, and 45° probes.

A-Scan Equipment

In an A-scan system the data is presented as a returned signal from the material under test. The data is presented on an oscilloscope. The horizontal base line of the oscilloscope screen indicates the elapsed time from left to right, and the vertical deflection shows signal amplitude. For a given velocity in the specimen, the sweep can be calibrated directly across the screen, in terms of distance or depth of penetration into the sample. Conversely, when the dimension (thickness) of the specimen is known, the sweep time may be used to determine ultrasonic velocities. The height of the indications or 'pips' represents the intensities of the reflected sound beams. These are used to determine the size of the discontinuity, and the depth, or distance to the discontinuity from any given surface the sound beam is either entering or reflecting back. The main advantage of this type of presentation is that it gives the amplitude that can be used to determine the size and position of the discontinuity.

B-Scan Equipment

A B-scan is especially useful when the distribution and shape of large discontinuities within a sample cross-section are of interest. In addition to the basic components of A-scan equipment, the B-scan provides the following functions:

1. It retains the image on the oscilloscope screen by use of a long persistence phosphor coating.
2. Deflection of the image-tracing spot on the oscilloscope screen is synchronized with the motion of the transducer along the sample.
3. Image-tracing spot intensity modulation or brightness is in proportion to the amplitude of the signals received.

C-Scan Equipment

C-scan equipment is intended to provide a permanent record of the test when high speed automatic scanning is used in ultrasonic testing. C-scan equipment displays the discontinuities in a plan view. It does not give the depth or orientation of the discontinuity.

TESTING PROCEDURE

Most of the testing is carried out as per written procedures for the specific work, which are based on the applicable code of construction. Hence the discussion here is of general application and specifics must be developed to meeting given work requirements.

Prior to testing with shear wave angle units, it is good practice and some codes mandate that the material is scanned with a longitudinal unit to ensure that the base material is free from such discontinuities that would interfere with shear wave evaluation of flaws.

Some of the basic rules of testing are listed below:

1. The sound path distance is basically limited to a specified limit generally up to 10 inches.
2. Three basic search unit angles are used – 70°, 60°, and 45° – as measured from a line normal to the test surface of the material.
3. It is assumed that any flaw will be normal to the test material surface and parallel to the weld axis. This flaw orientation would be the most serious direction for flaws in most welds.
4. The 70° search unit would provide the highest amplitude response from the type of flaw described above, followed by 60° and 45° search units. Hence the order of preference in use shall be the same.
5. The relative amplitude response from the flaw is in direct proportion to its effect on the integrity of the material and weld. The generally accepted diminishing order of flaw severity in welds and materials is given below.
   a. Cracks
   b. Incomplete fusion
   c. Incomplete penetration
   d. Inclusions (slag etc.)
   e. Porosity
6. Ultrasonic indications are evaluated on a decibel amplitude basis. Each indication to be evaluated is adjusted with the calibrated dB gain or attenuation control to produce a reference level height on the CRT, and the decibel setting number is recorded as indication level a.
7. Reference level b is obtained from a reflector in an approved calibration block. The reflector indication is maximized with search unit movement and then adjusted with the gain or attenuation control to produce a reference level indication. This decibel reading is the reference level.
8. Decibel attenuation factor c, used for weldment testing, is at the rate of two decibels per inch of sound path after the first inch. Example: A 6 inch sound path would produce an attenuation factor of $(6-1) \times 2 = 10$.

9. Decibel rating d, for flaw evaluation, is in accordance with the construction code requirement, with a gain control this is obtained by applying the equation a−b−c = d. However, for equipment with attenuation control, b−a−c = d.

Role of Coupling in Testing

The couplant material is used to maintain full contact of the transducer with the material surface. This allows the transfer of the sound wave. It also helps full coverage of the test surface during testing.

The coupling material should be hydraulic in nature, and have good wetting properties to cover the material's surface. The couplant materials used most often include water, oil, grease, glycerin, and cellulose gum powder mixed with water.

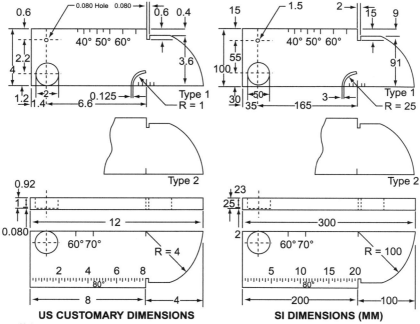

US CUSTOMARY DIMENSIONS   **SI DIMENSIONS (MM)**

Notes:
1. The dimensional tolerance between all surfaces involved in retaining or calibrating shall be within +0.005 inch (0.13 millimeter) of detailed dimension.
2. The surface finish of all surfaces to which sound is applied or reflected from shall have a maximum of 125 μin. r.m.s.
3. All materials shall be ASTM A36 or acoustically equivalent.
4. All holes shall have a smooth internal finish and shall be drilled 90 degrees to the material surface.
5. Degree lines and identification marking shall be indented into the material surface so that permanent orientation can be maintained.
6. Other approved reference blocks with slightly different dimensions or distance calibration slots are permissable.

FIGURE 3-6-2 Calibration block.

Cellulose gum powder is the most common material used for testing. It has significant advantages over the others, as it is low cost, its viscosity can be changed by addition of more water, it is not a slipping hazard, it does not form a contaminant film on the material surface and its residue is easily removable, hence it is not an obstacle for any further work like repairs involving welding etc.

Prior to testing, the equipment is checked for linearity and calibrated to cover the scope of the work. The calibration blocks used for calibrating the UT equipment before testing can be standard, or very specific to the task in hand. A typical IIW block is shown in Figure 3-6-2. Both US customary unit and metric unit calibration blocks are shown. Note that the block has a beam exit angle for correcting the angle of sound beam exit from the transducer. This is an important step in accurate testing. Most standards and code mandate that prior to the start of calibration and testing, the beam exit angle is re-established, and all future calculations are based on the correct angle of the beam path.

Eddy Current Testing

The eddy current testing process is also called electromagnetic testing. The method is based on the general principle that an electric current will flow in any conductor subjected to a changing magnetic field.

Depending on the type and thickness of material being tested, the testing frequencies vary from 50 Hz to 1 MHz. The method is used to check welds in magnetic and non-magnetic material, and is particularly useful in testing bars, billets, welded pipes, and tubes.

METHOD

In eddy current testing, an electric current – either an eddy or Foucault current – is induced in the test piece and the changes in that current are measured. The changes occur because of the presence of discontinuities. The test measures the resistivity caused by changes in chemical composition, crystal orientations, heat treatment hardness or discontinuities.

Applied Welding Engineering: Processes, Codes and Standards.

Acoustic Emission Testing (AET)

The application of acoustic emission testing (AET) includes detection of possible cracks in a vessel or structure. This is done on the principle that a sound structure would stop emitting signals once the load is reduced, and does not emit any further bursts until the previous load is exceeded. A growing crack emits an increasing signal as it is loaded, thus making it possible to find the crack's location.

AET has been successfully applied to welded pressure vessels and other welded structures during proof testing to determine the growth of cracks during pressurization. AET is often used as a surveillance method to monitor potential failures in vessels and structures whilst they are in service. As the vessel or the structure is loaded, and the cracks grow, it emits increasing decibels that are distinct from the rest of the acoustic signals. These signals are presented in a wide frequency spectrum along with ambient noise from many other sources. Transducers are strategically located on a structure and activated by the arriving signals. By use of proper filtering, the ambient noise in the composite signal is significantly reduced, and the source of significant signals is plotted by triangulation, based on the arrival time of these signals at different transducers.

ONGOING DEVELOPMENTS IN THE AET FIELD

The American Society of Mechanical Engineers (ASME) has developed a 'Standard for AE Examination During Application of Pressure'. Similarly The American Society for Testing Materials (ASTM) Committee E-7, that looks after non-destructive testing is engaged in developing a standard terminology for AE, transducers, method of applications, acoustic waveguides etc. This committee is also engaged in developing recommended practices for the calibration of a frequency response. This ASTM committee also deals with

AE instrumentations, such as recommended practice for performance of event counting and locating systems, and applications of AET.

The application of AET in weld testing is also possible, especially where delayed cracking is possible in the material.

FUTURE OF AET

AET equipment has evolved to a very refined state over a long development period. The evaluation of AET source and signals, however, needs further development and standardization. Use of AET during proof testing is of value to fabricators and users and it offers promise of wider application. AET monitoring can improve the production control of welding during fabrication.

Selection of NDE Methods Application and Limitations for Specific Cases: Case Study 3

We have noted in the text of the book that the NDE methods for evaluating a specific discontinuity can work as complementary to one another. Several NDE methods may be able to perform the same task with their limitations or advantages.

A planned analysis of the task in hand must be made for each test. The following variables are to be considered in developing an NDE procedure and choosing a method:

- Material manufacturing process
- Level of acceptability desired
- Accessibility of article
- Type and origin of discontinuity
- Cost of testing
- Available equipment for testing.

In the following paragraphs we will discuss some particular cases, looking at how an evaluation was made based on the metallurgical analysis of defects, and the limitations of various NDE methods.

Defects in Weldments

1. Heat Affected Zone (HAZ) cracking

This is a weld-related defect, related to both ferrous and non-ferrous welds. The cracks are open to the surface, and they are generally quite deep and often very tight. They are found to run parallel to the direction of weld in the heat affected zone of the parent metal.

Metallurgical evaluation: Hot cracking of the heat affected zone of weldments increases in severity with the increasing carbon equivalent.

NDE Method Selection Options

- Magnetic Particle Testing (MT). Very useful in detecting and evaluating the HAZ cracks in ferromagnetic material.
- Liquid Penetrant Testing (PT). Normally used for non-ferrous weldments.
- Radiographic Testing (RT). Not normally effective for detecting HAZ cracks.

- Ultrasonic Testing (UT). The process is utilized where a specialized application is developed. Strict acceptance standards are developed for evaluation of HAZ cracks. Sound beam deflection is dependent on factors like sharp versus rounded root radii and slope conditions.
- Eddy Current Testing (ET). Not normally used for detection of HAZ cracks, the process can be adopted to evaluate HAZ cracks in non-ferrous materials.

2. Surface Shrink Cracks

This weld-related defect is common to both ferrous and non-ferrous welds. The cracks are open to the surface on the face of the weld, fusion zone and base metal. The cracks are very small and tight, and they may run both parallel and transverse to the direction of weld.

Metallurgical evaluation: Shrink cracks are the result of improper and localized heating. Localized heating and cooling associated with welding sets up stresses that exceed the tensile strength of the material, causing it to crack. Restriction to expansion and contraction can also set up excessive stresses leading to cracks.

NDE Method Selection Options

- Magnetic Particle Testing (MT). Very useful in detecting and evaluating the shrinkage cracks in ferromagnetic material. Magnetizing may be done from different directions to get all cracks.
- Liquid Penetrant Testing (PT). Successfully used for detecting and evaluating shrinkage cracks in non-ferrous weldments.
- Radiographic Testing (RT). Not normally effective for detecting shrinkage cracks.
- Ultrasonic Testing (UT). The process is not utilized for evaluation of shrinkage cracks, as other NDE methods are capable of doing it at much lower cost.
- Eddy Current Testing (ET). Not normally used for detection of shrinkage cracks, the process can be used to evaluate shrinkage cracks in non-ferrous materials.

3. Inclusions

This is a weld defect associated with both ferrous and non-ferrous welds. The inclusions may be open to the surface or subsurface. Inclusions may be metallic or non-metallic. They may appear singly or linearly distributed or scattered in the weld.

Metallurgical evaluation: Metallic inclusions are particles of metal that have different density, like tungsten inclusions in the GTAW process. They may also be non-metallic like oxides, sulfides, slag, etc. trapped in the solidifying weld metal.

NDE Method Selection Options

- Radiographic Testing (RT). This method is most suitable to detect an inclusion in the weld. They appear as sharply defined, round, erratically shaped or elongated spots of lighter density on the radiographs. The dense material inclusion will appear lighter than the parent metal density on the film.
- Ultrasonic Testing (UT). Ultrasonic testing can also detect inclusions, such as planner discontinuity, but will not be able to define if that indication is an inclusion or some other discontinuity.
- Magnetic Particle Testing (MT). Not able to detect internal inclusions.

- Liquid Penetrant Testing (PT). Not able to detect internal inclusions.
- Eddy Current Testing (ET). Not normally used for detection of inclusions in welds except for welds on some applications on thin wall tubing.

4. Lack of Penetration

This is a weld defect that generally occurs at the root of the weld and runs parallel with it.

Metallurgical evaluation: Caused by the root face of the weld not reaching fusion temperature before the weld metal is deposited. It can also be caused by a fast weld rate, too large a welding rod, or too cold a bead.

NDE Method Selection Options

- Radiographic Testing (RT). This method is most suitable to detect an inclusion in the weld. They appear as sharply defined, linear indications running in the center of the weld. The indications may be long or intermittent. Good sensitivity and density of the film will ensure proper evaluation.
- Ultrasonic Testing (UT). Ultrasonic testing can also detect lack of fusion, it will appear on the scope as a definite break and appear as a planner discontinuity but will not be able to define if that indication is an inclusion or cracks or some other discontinuity.
- Magnetic Particle Testing (MT). Not able to detect internal inclusions.
- Liquid Penetrant Testing (PT). Not able to detect internal inclusions.
- Eddy Current Testing (ET). Not normally used for detection of lack of fusion in welds except for welds on some applications on thin wall tubing.

5. Gas Porosity

This is a weld defect associated with both ferrous and non-ferrous welds. The porosity may be open to the surface or subsurface. They may have an elongated, round teardrop shape. They may be clustered or scattered throughout the weld.

Metallurgical evaluation: Porosity in a weld is caused by gas entrapment in the molten metal. Too much moisture on the base metal or filler metal or improper cleaning or pre-heating may all cause and contribute to gas porosity.

NDE Method Selection Options

- Radiographic Testing (RT). This method is most suitable to detect porosity in the weld. They appear as oval shaped spots with smoother edges.
- Ultrasonic Testing (UT). Ultrasonic testing can also detect inclusions, as an area of separation.
- Magnetic Particle Testing (MT). Except for surface opening porosity, the MT method is not able to detect internal porosity in the welds.
- Liquid Penetrant Testing (PT). Except for surface open porosity the PT method is not able to detect internal porosity in the weld.
- Eddy Current Testing (ET). Not normally used for detection of porosity in welds except for welds on some applications on thin wall tubing.

Defects in Material Caused by Manufacturing and Service Processes

1. Burst

This is a defect associated with processing, found in ferrous and non-ferrous wrought material. It is manifest on either the internal or external surface, and appears as a straight or irregular cavity varying in size from wide open to very tight. It is often parallel to the grain of a forging, rolling or extrusion.

Metallurgical evaluation: Forging bursts are surface or internal ruptures caused by processing at too low a temperature, excessive working, or metal movement during rolling, or extrusion operations.

A burst has no spongy appearance, and therefore is distinguishable from pipe (pipe is also a type of defect), even if it occurs in the center.

Bursts are often large and are very seldom healed during subsequent working.

NDE Method Selection Options

- Ultrasonic Testing (UT). This method is normally used for detection and evaluation of burst. As the bursts are definite openings in the material exposed or hidden from the surface, they produce very sharp reflection on the scope. The UT method is capable of detecting variable degrees of burst and conditions not detected by other NDE methods. Limitations: Other defects like nicks, gouges, raised areas, sharp tool marks, tears, foreign material, on gas bubbles on the material may mask the result.
- Eddy Current Testing (ET). This method is restricted for use on wires, rods and other articles normally under 0.25 inch diameter, hence not normally used for mass defect evaluation of larger objects.
- Magnetic Particle Testing (MT). Applied to wrought ferromagnetic material in which the burst is open to the surface or has exposed to the surface. Results are limited to the surface opening or near surface exposure of burst.
- Liquid Penetrant Testing (PT). The method is limited to evaluation of surface opening bursts, hence the process is normally not used to detect and evaluate defects like burst.
- Radiography Testing (RT). The variables like direction of burst, close interfaces, wrought material geometry, discontinuity size, and material thickness may be restrictive in effective use of the method. This method is not normally used for detecting and evaluation of burst.

2. Cold Shuts

Cold shuts are produced during casting molten metal. They are inherent defects found in ferrous and non-ferrous cast material. They are open to the surface or may be subsurface, and appear as smooth indentations on the cast surface and resemble a forging lap. Cold shuts are the presence of interposing surface films of cold, sluggish metal, or any factor that prevents fusion of two meeting surfaces.

Metallurgical evaluation: Cold shuts result from the splashing, or surging of molten metal, or by interrupted pouring or meeting of two streams of metal coming from two different directions. They are also caused by solidification of one surface before other metal flows over it. They are often found in the casting molds that have several spurs or gates.

NDE Method Selection Options

- Liquid Penetrant Testing (PT). This is the primary method to evaluate surface cold shuts in both ferrous and non-ferrous materials. The indications appear as a smooth, regular, continuous or intermittent line. The PT method is best used on metal like nickel-based alloys, stainless steels and titanium where sulfur and chlorine are limited to 1 percent maximum. Sometimes the geometry of the casting may be restrictive in effective application of the PT method. Casting with blind surfaces, recesses, orifices and flanges, many cause difficulty in removal of excess penetrant or allow the developer to build up, thus masking the defect, and its evaluation.

- Magnetic Particle Testing (MT). More suitable for use to detect cold shuts in ferromagnetic material. Some duplex structure materials that have different magnetic retentivity may show indications that are in fact not cracks and thus mask or confuse with actual openings.
- Radiographic Testing (RT). Casting configuration is the limiting factor in evaluation of cold shuts in castings. If detected they appear as dark lines or bands of variable length and width with a definite smooth outline.
- Ultrasonic Testing (UT). Cast structure and article configuration do not, as a general rule, lend themselves to the UT test method. This method is not recommended.
- Eddy Current Testing (ET). Casting geometry and inherent material variables restrict use of this method. ET is not recommended for use to detect cold shuts in castings.

3. Hot Tears

Hot tears are the ruptures often associated with castings. They are the result of stresses developed due to the uneven cooling of material. They may be internal, or may appear near the surface as a ragged line of variable width, with numerous branches. The tears may be single or in groups.

Metallurgical evaluation: Hot tears are carks that are caused by the stresses developed by uneven cooling while the material is still in a brittle state. The tears are often at the point where the temperature gradient is abrupt, raising the possibility of more stresses. Such locations could be the junction of a thicker section with a relatively very thin section of the casting.

NDE Method Selection Options

- Radiographic Testing (RT). Since the material structure is cast and the tears may be internal, the method of choice is the radiography testing. However the sensitivity of the radiograph and also the orientation of the defect to the radiation beam could be the limiting factors.
- Magnetic Particle Testing (MT). Hot tears that are exposed to surface or near surface can be detected by the MT method. Magnetizing in different orientations may be a challenge in some cases. Pre-cleaning the surface is essential for successful results. A testing procedure that includes a combination of MT and RT must be developed in which MT may be used to screen out surface defects and RT may be used to detect internal tears.
- Liquid Penetrant Testing (PT). Non-ferrous material casting may be tested with PT in place of MT. The PT method will be able to detect only the surface opening hot tears. Absorption of penetrant in the openings may be a cause of concern for the next stage of processing,
- Ultrasonic Testing (UT). Cast structure and article configuration do not, as a general rule, lend themselves to the UT test method. This method is not recommended.
- Eddy Current Testing (ET). The metallurgical structure of casting as well as the complex configuration of the castings will not allow for proper detection or evaluation. This method is not recommended.

4. Micro Shrinkage

Micro shrinkage is specific to magnesium casting; this is an internal defect appearing as voids in the grain boundaries.

Metallurgical evaluation: Micro shrinkage is caused by the withdrawal of low melting constituents from the grain boundaries. Shrinkages occur when the metal is in a plastic or semi-molten state. If sufficient molten metal cannot flow into the molds while the casting is cooling, the shrinkages leave a void. These voids are identified by their appearance and by the time in the plastic stage when it occurs.

NDE Method Selection Options
- Radiographic Testing (RT). Radiography is the most effective method, and it is capable of determining the acceptable level of micro shrinkages.
  On the radiograph it appears as dark uneven patches, or an elongated swirl resembling streaks. They are indications of cavities in the grain boundaries.
- Liquid Penetrant Testing (PT). Used on the finished and machined products, micro shrinkages are normally not open to surface. On cut or machined portions the defects appear as hairline cracks or even as a large open porous indication.
- Eddy Current Testing (ET). Not recommended, as the casting configuration and type of discontinuity do not lend themselves to this method.
- Ultrasonic Testing (UT). Cast structure and article configuration do not, as a general rule, lend themselves to the UT test method. This method is not recommended.
- Magnetic Particle Testing (MT). Material is non-ferrous hence this method is not applicable.

5. Unfused porosity
This is an internal defect caused during processing, and associated with rolled, forged or extruded wrought aluminum. These are very thin fissures aligned with the flow of grains.

Metallurgical evaluation: This defect is often associated with cast ingots. During subsequent processes like rolling, forging or extrusion, the defects are flattened into very thin shapes. If the internal surfaces of these discontinuities are oxidized or have inclusions, then they do not fuse during the subsequent processes, and appear as very fine voids.

NDE Method Selection Options
- Ultrasonic Testing (UT). This method is extensively used as the most successful method to detect unfused porosity. The UT method determines the 3D location of the defect.
- Liquid Penetrant Testing (PT). This method is used on non-ferrous material where the defect is open to surface.
- Eddy Current Testing (ET). Normally not used.
- Radiographic Testing (RT). Thin openings are difficult to detect by this method, hence this method is normally not used.
- Magnetic Particle Testing (MT). Since the material is non-magnetic this method is not used.

6. Fillet Cracks
This is a service category defect of ferrous and non-ferrous materials. The defect is open to the surface at the junction of the bolt head to shank (fillet). The fillet crack originates outside and progresses inwards.

Metallurgical evaluation: The crack is caused at stress riser points like the sharp change in the diameter as is found in the head to shank transition in a bolt. During the service life of a bolt, repeated loading takes place in which the tensile load fluctuates in magnitude. These loads can cause fatigue failure starting at the point where the stress rises occur. Fatigue failures are surface phenomena; they start at the surface and propagate inwards.

NDE Method Selection Options
- Ultrasonic Testing (UT). Extensively used, a wide selection of transducers and equipment enable on the spot evaluation for fillet cracks.
- Liquid Penetrant Testing (PT). Frequently used during maintenance and overhaul processes. Indications are very sharp and clear.
- Magnetic Particle Testing (MT). Possibility to use is limited to ferromagnetic materials.
- Eddy Current Testing (ET). Not normally used.
- Radiographic Testing (RT). Not used.

7. Grinding Cracks
These are associated with heat-treated, case hardened, chrome plated or ceramic materials. This is a processing defect on the surface of a ferrous or non-ferrous metal. They appear at right angles to the direction of grinding. They appear as shallow and sharp at root, often resembling heat-treat cracks, but may appear in a group.

Metallurgical evaluation: These thermal cracks are often developed by local overheating of the surface being ground. Grinding of hardened surfaces often develops cracks. The use of proper coolant may prevent some of these cracks. Excessive feed during grinding or too heavy cut can cause such cracks.

NDE Method Selection Options
- Liquid Penetrant Testing (PT). Since these are surface opening defects they are easily detected by the PT method. The PT method is especially suitable for non-ferrous material. Since there are very tight and fine cracks they require extra dwell time for the penetrant.
- Magnetic Particle Testing (MT). The method is used on ferrous material with a similar effect as PT on non-ferrous materials. Since these cracks are often at right angles to the direction of grinding the orientation of magnetic field assumes importance.
- Eddy Current Testing (ET). Normally not used, however this method can be adopted for use on specific non-ferrous material.
- Ultrasonic Testing (UT). Not used as other methods are more effective and can be done at low cost.
- Radiographic Testing (RT). This method is less likely to detect these fine cracks, and so is not recommended.

8. Thread Cracks
This is a crack that develops in service. This is found on both ferrous and non-ferrous material.

Metallurgical evaluation: These cracks start at the root of the thread and are transgranular (transverse to the grain). They are fatigue cracks, often caused by cyclic stresses resulting from vibrations, flexing, or a combination of both at the stress risers at the root of the thread. They may start as small microscopic discontinuities and propagate in the direction of the applied stress.

NDE Method Selection Options

- Liquid Penetrant Testing (PT). PT is the most effective method for non-ferrous as well as ferrous materials. The fluorescent penetrant method is most effective. Cleaning of the thread is important as low surface tension liquids are not effective cleaners.
- Magnetic Particle Testing (MT). Equally effective method but limited to ferromagnetic material. The thread configuration may present non-relevant indications that may make interpretation difficult.
- Eddy Current Testing (ET). Not normally used, however specially adopted equipment for specialized use can be developed.
- Ultrasonic Testing (UT). The process is not suitable for this type of work.
- Radiographic Testing (RT). Not recommended. Other less costly and more effective methods are better suited for this kind of work.

9. Tubing Cracks

This is an inherent defect of non-ferrous tubes found in the internal (ID) surface of drawn tubes. These cracks are found parallel to the direction of grain flow.

Metallurgical evaluation: There can be several causes working individually or in combination that can form these cracks. The causes include; improper cold reduction, embedded foreign material during cold working, on insufficient heating during annealing.

NDE Method Selection Options

- Eddy Current Testing (ET). Eddy current method is most suitable to detect this type of defect in tubing. Small-bore tubing with low wall thickness can be covered by this method.
- Ultrasonic Testing (UT). A wide variety of equipment and transducers are available to use this method. The UT method may be limited by the temperature to which the transducers can work, if inspection is required on a hot-tube. Selection of sulfur-free couplant may be required on some material like high-nickel alloys.
- Radiographic Testing (RT). Discontinuity orientation and thickness are the limiting factors, so this method is not generally used.
- Liquid Penetrant Testing (PT). Not recommended.
- Magnetic Particle Testing (MT). Not applicable as the materials are non-magnetic.

10. Hydrogen Flake

These are process-related internal fissures that occur in ferrous materials, often found in steel forgings, billets and bars.

Metallurgical evaluation: The flakes appear as bright silvery areas on a fractured surface. If etched they appear as short discontinuities. In a machined surface they appear as hairline cracks. Flakes are very thin and are aligned parallel with the grain.

NDE Method Selection Options

- Ultrasonic Testing (UT). Most effective method and used extensively. The surface condition can determine if the material can be screened using either immersion or contact method. On an A-scan presentation, hydrogen flakes will appear as hash on the screen or loss of back wall reflection.
  Surface cleaning removal of loose scale dirt, oil, and grease is essential for proper inspection. Other irregularities arising from gouging, tool marks,

scarfing etc may mask the interpretation as they all cause loss of back wall reflection.

- Magnetic Particle Testing (MT). Possible to use on finished machined surfaces. Flakes appear as short discontinuities, and resemble chrome checks or hairline cracks.
- Liquid Penetrant Testing (PT). The discontinuities are very tight and small making it difficult to detect by this method, hence this method in not used.
- Eddy Current Testing (ET). The metallurgical structure of the material limits the use of this method.
- Radiographic Testing (RT). Not recommended, as the size and orientations present limitations to the RT method.

11. Lamination

Laminations can be found in both ferrous and non-ferrous wrought material, like forgings, extrusions, and rolled material. They can be found both on the surface as well as internally. They are extremely thin, and are generally aligned parallel to the work surface of the material. They may contain a thin film of oxide between the surfaces.

Metallurgical evaluation: Laminations are separations or weaknesses generally aligned parallel to the work surface of the material. They are caused when blister, pipe, seams, inclusions, or segregations are elongated and made directional by working. Laminations are flattened impurities that are extremely thin.

NDE Method Selection Options

- Ultrasonic Testing (UT). This is the most suitable method to detect most of the laminations. The use of various wave modes and angle probes in immersion or contact processes may be used to detect lamination on various geometrics of material. They appear as definite loss of back wall echo. For very thin sections through transmission and reflection techniques may be more effective. The UT method can plot out the size and depth of the lamination.
- Magnetic Particle Testing (MT). The surface opening ends of lamination may be detected by the MT method on ferromagnetic material. Lamination will appear like a straight line of intermittent linear indications.
- Liquid Penetrant Testing (PT). This method is suitable for use on non-ferrous material. The method has a similar effect and limitations as the MT method.
- Eddy Current Testing (ET). Not practical for use.
- Radiographic Testing (RT). Not recommended due to the orientation of lamination which is less likely to be resolved on RT film.

12. Laps and Seams

This is a surface defect caused during processing and is associated with both ferrous and non-ferrous materials.

Laps are found in wrought forgings, plates, tubing and bars. Laps appear on the surface as wavy lines. Usually laps are not very pronounced. They are often tightly adherent because they enter the surface at a small angle. Often their surface opening is smeared closed. Laps or rolled threads are the most common examples of such defects. Laps and seams appear on the surface as wavy lines; often quite deep and sometimes very tight, appearing as hairline cracks.

Seams are surface defects associated with rolled rods and tubing. They are often long, quite deep and sometimes very tight. They often occur in parallel fissures with the grain and at times are spiral.

Metallurgical evaluation: Seams originate from the blowholes, cracks, splits and tears discussed earlier. They are elongated in the direction of rolling or forging. The distance between adjacent interfaces of the discontinuity is very small.

Laps are similar to seams, and they can occur in any part of the material. They often result from improper rolling or sizing operations. During the processing of the material, corners may get folded over or an overfill may exist during sizing. These result in the material being flattened but not fused into the surface.

NDE Method Selection Options

- Magnetic Particle Testing (MT). This method is compatible with both ferrous and non-ferrous material. The indications may appear as straight, spiral, or slightly curved. They may be intermittent or continuous indications. Both laps and seams may appear as individual or a cluster indication. Magnetic build-up at laps and seams is very small, hence larger than usual magnetizing current is applied. In forgings the orientation of laps may lie in a plane that is nearly parallel to the surface.
- Liquid Penetrant Testing (PT). This method is applicable to non-ferrous material. Laps and seams may be too tight to be detected by the PT method. Application of fluorescent penetrant is the most preferred and successful method.
- Ultrasonic Testing (UT). A limited application of UT is possible for testing material prior to machining wrought material. Surface wave technique permits accurate evaluation of depth and area of the defect. Definite loss of echo between the inner faces of laps and seam is shown on the screen.
- Eddy Current Testing (ET). This method is used for evaluation of laps and seams in tubes and pipes. Material configuration and size are the limiting factors in extensive use of this method.
- Radiographic Testing (RT). The RT method is not recommended for detecting laps and seams.

13. Hydrogen Embrittlement

This is a surface discontinuity associated with ferrous materials. The defects are caused by service conditions and processing conditions. They are small, non-dimensional (interface) defects with no orientation or direction. They are often associated with materials that have been exposed to free hydrogen or subject to pickling or plating.

Metallurgical evaluation: Operations like pickling, electroplating, and cleaning prior to electroplating generates hydrogen at the surface of the material. This hydrogen penetrates the surface of the material creating immediate or delayed cracking – called hydrogen embrittlement.

NDE Method Selection Options

- Magnetic Particle Testing (MT). Indications often appear as a fractured pattern, hydrogen embrittlement cracks are randomly oriented and can be often aligned with magnetic field. MT testing must be carried out both before and after plating operations.

- Liquid Penetrant Testing (PT). Not normally used for hydrogen embrittlement crack detection.
- Ultrasonic Testing (UT). This method is not suitable to detect this type of cracks.
- Eddy Current Testing (ET). Not recommended for detecting these cracks.
- Radiographic Testing (RT). Not capable of detecting hydrogen embrittlement cracks.

Corrosion Related Defects and their Detection

1. Intergranular Corrosion (IGC)

This is a defect that occurs externally as well as internally. IGC is a service-condition-related defect that can also develop during service due to faulty fabrication processes. It is exclusively a non-ferrous material defect, appearing as a series of small micro-openings with no definite pattern, either singly or in groups. If IGC appears on the surface then in most cases it is too late to save the material. As the name suggests it is intergranular and follows the grain boundaries of the material.

Metallurgical evaluation: Factors that contribute to IGC are:

(a) Unstabilized austenitic stainless steel.

(b) Improper heat treatment generally solution annealing.

Any of these two conditions can cause IGC during the service life of the material.

NDE Method Selection Options

- Liquid Penetrant Testing (PT). This is the best-suited method for detection of IGC in material. Selection of chlorine-free and chlorine-emitting cleaning product is essential.
- Radiographic Testing (RT). In more advanced conditions of IGC, radiography may be able to detect the cracks. Correct location of the defect on the film and sensitivity of the film may be a problem.
- Eddy Current Testing (ET). ET method can be used as a screening method of tubes and pipes.
- Ultrasonic Testing (UT). Not normally used as the method is not suitable for detecting such cracks.
- Magnetic Particle Testing (MT). Not applicable as material is non-magnetic.

2. Stress Corrosion Cracks (SCC)

This is a service-condition-related defect occurring in both ferrous and non-ferrous materials. The cracks are often very deep and often follow the grain flow, however transverse cracks are also possible.

Metallurgical evaluation: For stress corrosion to occur the following three phenomena must co-exist:

(1) A sustained static tensile stress

(2) Presence of corrosive environment

(3) Use of material that is susceptible to this type of failure.

Stress corrosion is most likely to occur at high rather than low levels of stress. The stresses include residual and applied stresses.

NDE Method Selection Options

- Liquid Penetrant Testing (PT). Most common method to use for the detection of stress corrosion cracks, especially on non-ferrous materials.

- Magnetic Particle Testing (MT). MT is mostly used on ferrous material, for similar results as the PT method.
- Eddy Current Testing (ET). Eddy current testing is capable of resolving some SCC indications depending on whether the configuration is compatible with equipment.
- Ultrasonic Testing (UT). Since the discontinuity orientation is mostly perpendicular to the surface, surface wave technique may be required but that would limit the results and increase the cost of inspection.
- Radiographic Testing (RT). Not suitable to detect SCC in most cases.

Ferrite Testing

EFFECT OF FERRITE IN AUSTENITIC WELDS

Weld deposits that are fully austenitic have a tendency to develop small fissures, even under conditions of minimum stress. These small fissures tend to be located transverse to the weld fusion line, in weld passes and base material that were reheated to near the melting point of the material during subsequent weld passes.

The effect of these micro-fissures on the performance of the weldments is not well established, because it is clear that the tough austenitic matrix blunts the progression of these fissures. Several fissure-containing weldments are known to exist and have performed satisfactorily under severe conditions. However the tendency to develop fissures goes hand in hand with the development of possible larger cracks that could cause catastrophic failure. This has motivated the development of a process that should reduce or eliminate the chances of micro-fissures occurring.

It is well established that the presence of a small fraction of magnetic delta ferrite phase in an otherwise austenitic weld deposit has a pronounced influence in the prevention of both centerline cracking and fissuring. Both the presence and amount of delta ferrite in the as-welded material is largely influenced and dependent on the composition of the weld metal. The balance between the elements which form ferrite and austenite is the key to the presence of δ ferrite. The most common ferrite and austenite forming elements are listed in Table 3-9-1. The subject is discussed in detail with the help of DeLong and Schaeffler's diagrams in Section 2, Welding Metallurgy.

Having learned about the advantages of having δ ferrite in a weld, we must note that an excessive amount of ferrite is also not desirable in welds, as it lowers the ductility and toughness of the weld metal. Delta (δ) ferrite is also preferentially attacked by corrosive environments. It is also attacked in its sensitizing range of temperature – that is between 800°F and 1,600°F (425°C to 870°C). In

TABLE 3-9-1

Ferrite forming elements	Austenite forming elements
Chromium	Nickel
Silicon	Manganese
Columbium	Carbon
Molybdenum	Nitrogen

this temperature range, ferrite tends to partly transform to a brittle intermetallic compound called sigma phase that severely embrittles the weldments.

This requires that the level of ferrite in austenitic weldments must be controlled. This is done by metallographic examination of a specimen and calculation of the percent volume ferrite present. This method is not very favored for several reasons; one being that the distribution of ferrite in weld metal is not uniform and the sample preparation procedure method is very tedious. The impact of a faulty result is too serious to risk such a method.

Chemical analysis of weld metal using constitution diagrams like Schaeffler and DeLong's is very common and popular in use. The result of this method depends on the accuracy of the analysis.

Since ferrite is magnetic, it can be measured by magnetic responses of the austenitic material. This measurement is reproducible in laboratories if standard calibrated equipment is used. Most instruments are able to convert a magnetic force reading to a standard ferrite measurement.

Ferrite content has traditionally been expressed as a percent volume of the weld metal until the ferrite number (FN) was recommended. Since there is no agreement in laboratories on reporting of the absolute ferrite percentage, an arbitrary FN system has been developed for reporting. The American Welding Society (AWS) has developed a scale and it is described in AWS 4.2-74, "Standard Procedure for Calibrating Magnetic Instruments to Measure the Delta Ferrite Content of Austenitic Stainless Steel Weld Metal". The FN scale, although arbitrary, approximates the true volume percentage of ferrite at least up to 10 FN.

The ferrite content that is recommended for weld filler metal is usually between 3 and 20%. A minimum of 3% ferrite is desirable to avoid microfissuring in welds. The upper limit of 20% ferrite is required when needed to offset dilution losses in weld metal. Delta (δ) ferrite verification can be made by tests on undiluted weld deposits using magnetic measuring devices. AWS 5.4 details the procedure for preparation of pads for ferrite measurement.

The testing of ferrite in a laboratory setting is very different from the portable ferrite indicators used in field sites. In some cases the details and results can vary significantly between the two methods.

Pressure Testing

PURPOSE

Pressure tests are carried out to induce a pre-determined stress level in a sample, by pressurizing the equipment and observing the results.

Leak tests are carried out to determine the extent of a flaw's depth; that is to ensure if the flaws in the material or weld extend to the surface of the material. The test can either be simple, such as pressurizing to a relatively low pressure to create a pressure differential and visually inspecting for leaks, or sophisticated methods using electronic equipment like acoustic emission etc., to detect possible cracks and their growth over time.

METHOD

In its simplest form, the test is carried out by increasing the internal pressure, and creating a pressure differential with ambient pressure. This allows liquid to flow out of possible openings; that is the leak. This is inspected by visual methods.

If fluorescent dyes are used, the inspection is carried out in a dark place, and the use of black (ultraviolet) light is required.

If the test media used is air or gas, then the leaks are located by looking for the rising bubbles from the test surface. An emulsion of mild soap and water is applied on the test surface to facilitate this.

Sometimes pressurized components are immersed in water to detect leak locations.

Applied Welding Engineering: Processes, Codes and Standards.

Test Medium

The media used for leak testing can be any liquid that is non-hazardous to personnel or the environment. Water is the most common medium for testing, although light oils can also be used. When water or any other liquid is used as a medium for testing, it is often called hydro-testing. A typical pipeline hydro-testing is discussed at the end of this chapter.

Some tests require the use of air or gas. A gas test is very sensitive for detecting small leaks, but both air and gas as test media must be used with utmost care, as they have the inherent disadvantage of explosive effects in the event of a failure.

Sensitivity of the Test

The test can be made more effective and sensitive by use of lighter chemicals, gases or by the addition of fluorescent dyes to the water.

The degree of sensitivity is adjusted for the required degree of flaw detection and relative degree of cost and risk involved in testing.

Proof Testing

Proof testing is a relatively high-pressure test compared to the leak test. A proof test is carried out to determine if the system can withstand applicable service loadings without failure or permanent deformation in the part.

The proof test is generally designed to subject the material to stresses above those that the equipment is expected to carry in service. Such service stress is always below the yield strength of material of construction. In this respect, also read the discussion on hydro-testing of pipeline at the end of this chapter.

The methods of testing vary according to the specific design and requirements of service. Mostly, the proof test is applied in conjunction with visual inspection. These specific requirements are generally specified in the project specification or dictated by the code of construction.

The test medium is usually water (hydrostatic test) or air (pneumatic test) – which is ordinarily used for relatively low pressure testing, because of the inherent safety associated with compressed air.

Hydrostatic testing has been used to determine and verify pipeline integrity. A lot of information can be obtained through this verification process, however it is essential to identify the limits of the test process and the obtainable results. There are several types of flaws that can be detected by hydrostatic testing such as:

- Existing flaws in the material
- Stress corrosion cracking (SCC) and actual mechanical properties of the pipe

- Active corrosion cells
- Localized hard spots that may cause failure in the presence of hydrogen etc.

There are some other flaws that cannot be detected by hydrostatic testing, for example sub-critical material flaws, but the test has a profound impact on the post-test behavior of these flaws.

The process of proof testing involves the following steps:

- Pressurizing the vessel or pipe with water to a stress level that is above the design pressure but below the yield strength of the material of construction
- Holding the pressure for the required time
- Monitoring for a drop in pressure
- Inspecting the object for any leak while it is under the test pressure
- Controlled depressurizing.

Vessels or pipe sections that are pressurized with air are often inspected by application of emulsified soap and inspected for rising bubbles from the leak locations. Sometimes pressurized components are immersed in water to detect leaks.

PRACTICAL APPLICATION OF HYDROSTATIC TESTING

As already stated, hydrostatic testing is used to determine and verify pipeline integrity. There are several data that can be obtained through this verification process; however it is essential to identify the limits of the test process and obtainable results. We have seen that there are several types of flaw that can be detected by hydrostatic testing, whereas some other flaws cannot be detected by this method of testing.

Given that the test will play a significant role in the non-destructive evaluation of a pipeline, it is important to utilize the test pressure judiciously. The maximum test pressure should be so designed to provide a sufficient gap between the test pressure and the maximum operating pressure (MOP), in other words:

The maximum test pressure should be $> ^{Gap}$ MOP.

This also pre-supposes that after the test the surviving flaws in the pipeline will not grow when the line is placed in service at the maintained operating pressure. When setting the maximum test pressure, it is important to know the effect of pressure on the growth of the defect, and how this growth would be affected by pressure over time. These defects are often referred to as sub-critical defects, because they will not fail during a one-time high pressure test, but they would fail at a lower pressure if this is held for a longer time. The size of discontinuity in the sub-critical group will be those that would fail – independent of time – at about 105% of the hold pressure. This implies that the maximum test pressure should be set to at least 5% to 10% above the MOP in order to avoid the growth of sub-critical discontinuities during the life of the pipeline.

The phenomenon of pressure reversal occurs when a defect survives a higher hydrostatic test pressure but it fails at lower pressure in a subsequent repressurization. One of the many factors that work to bring this phenomenon about is the creep-like growth of sub-critical discontinuities over time at the lower pressure. The reduction in wall thickness in effect reduces the discontinuity depth to the material thickness (d/t) ratio. This increase in d/t ratio reduces the ligament of the adjoining defects that in effect reduces the required stress to propagate the discontinuity. The other factor affecting pressure reversal is damage to the crack tip opening as it is subject to some compressive force leading the crack tip to force close. This facilitates the growth of the crack upon repressurization to a much lower pressure. Hence, if such pressure cycling is part of the design, then pressure reversal is a point of consideration.

When a pipeline is designed to operate at a certain MOP, it must be tested to ensure that it is structurally sound, and can safely withstand the internal pressure, before being put into service. Generally, gas pipelines are hydro-tested by filling a test section of pipe with water, raising the pressure up to a value higher than MOP, and holding it at this pressure for a period of 4 to 8 hours. The magnitude of the test pressure is specified by code, and is usually 125% of the operating pressure. Thus a pipeline designed to operate continuously at 1,000 psig will be hydrostatically tested to a minimum pressure of 1,250 psig.

Let's consider a pipeline NPS 32, with 12.7 mm (0.500 inch) wall thickness, constructed of API 5L X 70 pipe. Using a temperature derating factor of 1.00, we calculate the MOP of this pipeline from the following:

$$P = \{2 \times t \times SMYS \times 1 \times factor \,(class \, 1) \times 1\} / D \tag{1}$$

Substituting the values:

$$2 \times 0.5 \times 70\,000 \times 1 \times 0.72 \times 1 / 32 = 1,575 \text{ psig}$$

If the same pipeline had a design factor of 0.8 this pressure will be 1,750 psig.

If the fittings are of ANSI 600, then the maximum test pressure will be $(1.25 \times 1,440^{(ASME\,B\,16.5)})$ 1,800 psig.

If however ANSI 900 fittings were chosen, the pressure would be $(1.25 \times 2,220^{(ASME\,B\,16.5)})$ 2,775 psig.

If the selected design factor is for a class one location, then the factor would be 0.72. In this case the test would result in the hoop reaching 72% of the specified minimum yield strength (SMYS) of the material. Testing at 125% of MOP will result in the hoop stress in the pipe reaching a value of $1.25 \times 0.72 = 0.90$, or 90% of SMYS. Thus, in hydro-testing the pipe at 1.25 times the operating pressure, we are stressing the pipe material to 90% of its yield strength – that is 50,400 psi (factor $0.72 \times 70,000$).

Alternatively, however, we can use a design factor of 0.8 as is now often used, and allowed by several industry codes. Testing at 125% of MOP will raise the hoop stress in the pipe to $1.25 \times 0.8 = 1$. The hoop stress would reach 100% of the SMYS. So, at the test pressure of 1,800 psig the S_h (hoop stress) will be 56,000 psi ($0.8 \times 70,000$).

This will be acceptable if class 600 fittings where the limiting components in the system. But if class 900 fittings were taken into account as limiting the test pressure, then the maximum test pressure would be ($1.25 \times 2,220$) 2,775 psig, and the resulting stress would be 88,800 psi, which is far above the maximum yield stress of API 5L X 70 PSL-2 material.

Critical Flaw Size

Using a test pressure at 100% of the material yield strength has some important pre-conditions attached to it. Such a test pressure would require that the acceptable defect size be re-assessed. All being equal, a higher design factor, resulting in a thinner wall (d/t), will lead to a reduction in the critical dimensions of both surface and through-wall defects. Critical surface flaw sizes at design factors of 0.80 and 0.72 flow stress dependent will also be dependent on the acceptable Charpy energy for the material and weld.

As stated in the above discussions, the increase in d/t ratio in effect reduces the ligament of the adjoining defects, which reduces the stress required to propagate the discontinuity. Critical through-wall flaw lengths are also factors to be assessed. While there is a modest reduction in critical flaw length, it still indicates very acceptable flaw tolerance for any practical depth, and the reduction will have negligible influence in the context of integrity management. Note that flaws deeper than about 70% of wall thickness will fail as stable leaks in both cases. This statement implies that mere radiography of the pipe welds (both field and mill welds) will not suffice; automatic ultrasonic test (AUT) of the welds will be better suited to properly determining the size of the planar defects in the welds. Similarly the use of AUT for assessing the flaws in the pipe body will have to be more stringent than usual.

Codes and Standards

Chapter 1

Introduction

There are many codes and standards of interest to a welding engineer, and their relevance depends on what type of project is involved. This section is an introduction to some of the codes and specifications that are in common use in general engineering and the oil, gas, and petrochemical industries in particular. They also find application in various other industries.

Historically, incidents of pressure vessel incidents, bridge failures, and, notably, steam boiler industrial accidents led to the development of regulatory codes. These codes relate to the safety of men and material through design and inspection, with awareness of the damage these failures and accidents can cause to the environment. The emphasis on safety has been comprehensively integrated with concerns over damage to the environment.

At the end of this section, there is a list of some very commonly used specifications and their sources. Interested readers may obtain them from the sources indicated.

It may however be noted that the bodies that issue industrial specifications and the national codes have very exhaustive lists of fields that they address. As a result, the number of topics they address and the specifications they issue are exhaustive too, so it is not possible for a book to cover even a small fraction of them. In this book an attempt is made to introduce some of them so that readers may explore any specifications of interest to them by using the issuing body's website, and refer to any specific publication that may be of specific interest.

Codes, Specifications and Standards

The American National Standards Institute (ANSI) www.ansi.org and The American Society of Mechanical Engineers (ASME) www.asme.org are the governing organizations for many documents relating to material selection. Similarly, The American Society for Testing and Materials (ASTM) www.astm.org is the primary source of specifications relating to various materials (metals and non-metals), test methods and procedures, including for example, corrosion-resistant materials and various kinds of mechanical and corrosion tests.

The specifications issued by various institutions, including both ASME and ASTM, are adopted as part of the American National Specification Institute (ANSI).

Specifications are documents legally prescribing certain requirements regarding composition, mode of manufacture and physical and mechanical properties. Standards are documents representing a voluntary consensus.

However, when these specifications are adopted by a regulatory authority they become a code, such as the Boiler and Pressure Vessel Code (ASME), the National Building Code etc.

The National Association of Corrosion Engineers (NACE) www.nace.org has committees that write standards and exchange information in specific industries or particular areas of concern. NACE standards consist of recommended practices, materials requirements, and test methods for a variety of corrosion control or material selection challenges.

In the following paragraphs, we will try to give an introductory perspective on some of the key institutions through their historical development.

AMERICAN SOCIETY OF MECHANICAL ENGINEERS (ASME)

Located at Three Park Avenue, New York, NY 10016-5990, (www.asme.org), an introduction to ASME would be incomplete without the following words of Bruce Sinclair, an American historian.

"The Society's history helps to reveal the outlines and consequences of a complex technological information-processing system. It is an article of faith that Americans are inventive people. But besides machines, they also created a welter of interrelated institutions to translate technical knowledge into industrial practice, and that may have been one of the country's most successful inventions."

Background and History

The 19th century America through the eyes of a mechanical engineer: ASME was founded in 1880 by prominent mechanical engineers, led by Alexander Lyman Holley (1832–1882), Henry Rossiter Worthington (1817–1880), and John Edson Sweet (1832–1916). Holley chaired the first meeting, which was held in the New York editorial offices of the 'American Machinist' on February 16, 1880, with thirty people in attendance. On April 7, a formal organizational meeting was held at the Stevens Institute of Technology, Hoboken, New Jersey, with about eighty engineers, industrialists, educators, technical journalists, designers, shipbuilders, military engineers, and inventors present.

The later part of 19th century witnessed the widespread establishment of schools and institutions in engineering. Engineers of the day moved easily between the disciplines of civil, industrial, mechanical and mining engineering, with less distinction among them. Many groups sought to create organizations of specialized professional standing. But for mechanical engineers, none were devoted to machine design, power generation and industrial processes to a degree that was capable of projecting a broader national or international role to advance technical knowledge, or systematically facilitate a flow of information from research to practical application.

The Institution of Chartered Mechanical Engineers had been successfully established in England 33 years earlier in 1847. In the United States, the American Society of Civil Engineers had been active since 1852, and the American Institute of Mining Engineers had been organized in 1871. Holley had been vice-president of one and president of the other.

Mechanical engineers practiced in industries such as railroad transportation, machine tools, steel making, and pumping. In 1880 there were 85 engineering colleges throughout the United States, most of them offering a full mechanical engineering curriculum, with students gaining a degree in mechanical engineering.

The first annual meeting of ASME was held in early November 1880. Robert H. Thurston, professor of mechanical engineering at Stevens Institute and later Cornell, was the first president. Thurston is credited with establishing the first model mechanical engineering curriculum and laboratory.

This was an era of steel power that drove the technology of the day: locomotives, ships, factory machinery, and mine equipment. The Corliss engine and the Babcock & Wilcox water-tube boiler were in their prime. The first real US central power plant – Thomas Edison's Pearl Street Station in New York City – ushered in the era of great electric utilities in 1882. The internal combustion engine was not far from application. Conglomerates such as US Steel were formed. Industrial research laboratories, such as those at General Electric, du Pont, and Eastman Kodak, proliferated.

The early 20th century: ASME began its research activities in 1909, devoting its efforts to areas such as steam tables, the properties of gases, the properties of metals, the effect of temperature on strength of materials, fluid meters, orifice coefficients, etc.

Since its inception, ASME has led in the development of technical standards, beginning with the screw thread and now numbering more than 600 specifications. The society is best known, however, for improving the safety of equipment, especially boilers. This is for good reason. If we review the archives of engineering we find that between 1870 and 1910, at least 10,000 boiler explosions in North America were recorded. By 1910 the rate jumped to 1,300 to 1,400 a year. Some were spectacular accidents that aroused public outcries for remedial action. A Boiler Code Committee was formed in 1911 that led to the Boiler Code being published in 1914–15 and later incorporated into the laws of most US states and territories, and those of Canadian provinces.

By 1930, fifty years after ASME was founded, the Society had grown to 20,000 members, though its influence on American workers is far greater. Just as the nineteenth-century railroad created towns and cities along its paths, and its interlocking schedules led to establishment of present time zones, twentieth-century ASME leaders, such as Henry Robinson Towne, Fredrick W. Taylor, Frederick Halsey, Henry L. Gantt, James M. Dodge, and Frank and Lillian Gilbreth have pioneered management practices that have brought worldwide reform and innovation to labor-management relations. Precision machining, mass production, and commercial transportation opened the nation and then the world to American enterprise.

The diversity of mechanical engineering can be seen in ASME's 36 technical divisions (plus one subdivision) and 3 institutes. Today's structure of technical divisions was established in 1920, when eight divisions were created.

The primary divisions were:

1. Aerospace
2. Fuels
3. Management
4. Materials
5. Materials Handling Engineering
6. Power
7. Production Engineering, and
8. Rail Transportation.

Two more were added the next year:

1. Internal Combustion Engine and
2. Textile Industries.

The most recent addition, in June 1996, is the division called:

1. Information Storage and Processing Systems Division.

Present Day ASME

Currently, ASME is a worldwide engineering society focused on technical, educational and research issues. It has 125,000 members and conducts one of the world's largest technical publishing operations. It holds some 30 technical conferences and 200 professional development courses each year, and sets many industrial and manufacturing standards.

Under the sponsorship of the ASME, the Boiler and Pressure Vessel Committee establishes rules of safety governing the design, fabrication, and inspection during construction of boilers and pressure vessels. ASME is made up of various committees. There are several codes developed by sub-committees. The fabrication codes include:

List of all Twelve ASME Boiler and Pressure Vessels Codes

Section I, Power Boilers
Section III, Nuclear Codes
Section IV, Heating Boilers
Section VIII, Division 1 and 2, Pressure Vessels
Section X, Fiberglass Reinforced Vessels.

There are also reference codes issued to support the construction codes. These include:

Section II, Materials
    Part A, Ferrous Materials
    Part B, Non Ferrous Materials
    Part C, Welding Materials
    Part D, Materials Properties
Section V, NDE
Section IX, Welding Qualifications

The following is a brief discussion of some key ASME specifications and their role in the design and construction of pressure vessels. The key differences between ASME section VIII division 1, division 2, and division 3 are given in Table 4-2-1.

ASME Section VIII, Division 1 (Pressure Vessels)

The organization of Section VIII, Division 1 is as follows:

1. **Subsection A** General requirements
    Part UG   General requirements for all methods of construction and all requirements
2. **Subsection B** Methods of fabrication
    Part UW   Fabricated by welding
    Part UF   Fabricated by forging
    Part UB   Fabricated by brazing
3. **Subsection C** Materials
    Part UNC   Carbon steel
    Part UNF   Nonferrous
    Part UHA   High alloy steel
    Part UCI   Cast iron
    Part UCL   Cladding and weld overlay
    Part UCD   Cast ductile iron
    Part UHT   Heat-treated ferritic steels
    Part ULW   Layered construction
    Part ULT   Low-temperature materials
4. **Mandatory Appendices**   (Indicated by numbers)
5. **Non-mandatory Appendices**   (Indicated by letters)

ASME Code for Pressure Piping

ASME Code committee B31 has developed codes for pressure piping. They include B31.1 for power piping and B31.3 for chemical piping. In addition, the B31 committee publishes a supplement on corrosion, B31G, entitled Manual for Determining the Remaining Strength of Corroded Pipelines.

TABLE 4-2-1 A Brief Discussion on ASME Section VIII Div. 1, Div. 2 and Div. 3

Published	Division 1	Division 2	Division 3
	1940	1968	1997
Structure of Code			
Organization	General, Construction Type & Material U, UG, UW, UF, UB, UCS, UNF, UCI, UCL, UCD, UHT,UL.	General, Material, Design, Fabrication and others AG, AM, AD, AF, AR, AI, AT, AS.	Similar to Division 2 KG, KM, KD, KF, KR, KE, KT, KS.
Design			
Pressure Limits	Normally up to 3000 psig.	No limits either way, usually 600+ psig, no limit.	Normally from 10,000 psig.
Design Factor	Design factor 3.5 on tensile (4* used previously) and other yield and temperature considerations.	Design factor of 3 on tensile (lower factor under review) and other yield and temperature considerations.	Yield based with reduction factor for yield to tensile ratio less than 0.7.
Design Rules	Membrane - maximum stress generally elastic analysis. Very detailed design rules with quality (joint efficiency) factors. Little stress analysis required; pure membrane without consideration of discontinuities controlling stress concentration to a safety factor of 3.5 or higher.	Shell of revolution - max. shear stress generally elastic analysis. Membrane + bending. Fairly detailed design rules. In addition to the design rules, discontinuities, fatigue and other stress analysis considerations may be required unless exempted and guidance provided for in Appendix 4, 5 and 6.	Maximum shear stress elastic/plastic analyses and more. Some design rules provided; fatigue analysis required; fracture mechanics evaluation required unless proven leak-before-burst. Residual stresses become significant and may be positive factors (e.g. autofrettage).
Experimental Stress Analysis	Normally not required.	Introduced and may be made mandatory in future.	Experimental design verification but may be exempted.

(Continued)

TABLE 4-2-1 (Continued)

Published	Division 1	Division 2	Division 3
	1940	**1968**	**1997**

Material and Testing

Impact Testing	Few restrictions on materials; impact required unless exempted; extensive exemptions under UG-20, UCS 66/67.	More restrictions on materials; impact required in general with similar rules as Division 1.	Even more restrictive than Division 2 with different requirements. Fracture toughness testing requirement for fracture mechanics evaluation crack tip opening displacement (CTOD) testing and establishment of KIc and/or JIc values.
NDE Requirements	NDE requirements may be exempted through increased design factor.	More stringent NDE requirements; extensive use of RT as well as UT, MT and PT.	Even more restrictive than Division 2; UT used for all butt welds, RT otherwise, extensive use of PT and MT.

Method of Construction

Welding and fabrication	Different types with butt welds and others.	Extensive use/ requirement of butt welds and full penetration welds includes nonpressure attachments.	Butt welds and other construction methods such as threaded, layered, wire-wound, interlocking strip wound etc.

Common Features: Jurisdictional requirements may be in addition to the Code requirements.
Mandatory Manufacturer's Quality Control System Implementation and Audit Requirements are imposed.
Code Stamp Authorization through ASME Accreditation and Authorization process.
Authorized Inspection Agency in accordance with QAI of ASME.
Authorized Inspector with Jurisdictional approval and certification may be additional requirements.
Manufacturer is held accountable for Code Stamp Application and full Code Compliance.
NDE Personnel qualification to SNT-TC-1A
Note: *This brief comparative table is presented for basic introduction and discussion and does not represent the opinion of the ASME or the ASME Boiler and Pressure Vessel Code Committees. Readers are strongly advised to consult with the ASME Code Section VIII Divisions 1, 2 & 3 and jurisdictions requirements for more details of the subject.*

Some of the other pipeline related design and construction specifications are given in Table 4-2-2 below.

ASME Section V

ASME Code Section V is the reference code that contains the requirements for non-destructive examinations that are code requirements, and are referenced and required by other codes. It totals 30 articles, some of which are not titled, as listed in Table 4-2-3.

This section has two subsections. Subsection A describes the methods of non-destructive examinations to be used if referenced by other code sections, and subsection B lists the standards covering non-destructive examination methods that have been accepted as standards.

The National Board

The National Board (NB) is an organization made up of law enforcement officials in the United States and Canada. They administer and enforce boiler and pressure vessel laws within their jurisdiction. The NB also standardizes inspector's qualifications, and issues commissions to authorized inspectors who successfully pass the examinations.

Authorized Inspection Agencies are the organizations that employ Authorized Inspectors. The Agency may be either the jurisdiction charged with the enforcement of the boiler or pressure vessel laws, or an insurance company authorized to write boiler and pressure vessel insurance within a jurisdiction.

The National Board of Boiler and Pressure Vessel Inspectors is an organization comprised of Chief Inspectors for the states, cities and territories of the United States, the provinces and territories of Canada, and Mexico. It is organized for the purpose of promoting greater safety for life and property, by securing

TABLE 4-2-2 Currently the B 31 Committee Handles the Following Specifications

	Number	Title
1	B 31.1	Power Piping
2	B 31.2	Fuel Gas Piping
3	B 31.3	Petroleum refinery Piping
4	B 31.4	Liquid Petroleum Transportation Piping system
5	B 31.5	Refrigeration Piping
6	B 31.8	Gas Transportation and Distribution Piping system

TABLE 4-2-3 ASME Section V Articles

Article	Subject - Title
1	General Requirements
2	Radiographic Examination
3	Nil
4	Ultrasonic Examination Methods for in-service Inspection
5	Ultrasonic Examination Methods for Materials and Fabrication
6	Liquid Peneterant Examination
7	Magnetic Particle Examination
8	EDDY Current Examination of Tubular Products
9	Visual Examination
10	Leak Testing
11	Acoustic Emission Examination of Fiber-Reinforced Plastic Vessels
12	Acoustic Emission Examination of Metallic Vessels During Pressure Testing
13	Continuous Acoustic Emission Monitoring
14	NA
15	NA
16	NA
17	NA
18	NA
19	NA
20	NA
21	NA
22	Radiographic Standards
23	Ultrasonic Standards
24	Liquid Peneterant Standards
25	Magnetic Particle Standards
26	Eddy Current Standards
27	Leak testing Standards
28	Visual examination Standards
29	Acoustic Emission Standards
30	Terminology for Nondestructive examinations Standard

concerted action and maintaining uniformity in the construction, installation, inspection, repair and alteration of pressure-retaining items. This assures acceptance and interchangeability amongst the jurisdictional authorities that are responsible for the administration and enforcement of various codes and standards.

The National Board Inspection Code (NBIC)

The purpose of the National Board Inspection Code (NBIC) www.nationalboard.org is to maintain the integrity of pressure-retaining items after they have been placed in service, by providing rules for inspection, repair and alteration. This ensures that these objects may continue to be safely used.

The NBIC intends to provide guidance to jurisdictional authorities, inspectors, users and organizations performing repairs and alterations. This encourages the uniform administration of the rules pertaining to pressure-retaining items.

AMERICAN PETROLEUM INSTITUTE

The American Petroleum Institute (API) www.api.org, located at 1220 L Street, Northwest, Washington, DC 20005, is a trade association representing the entire petrochemical industry. The chemical process industry adopted the API standards for chemical process tanks and vessels. API began in 1919, evolving from the need to standardize engineering practices and specifications for drilling and production equipment. API has developed more than 500 standards related to the oil and gas industry. API requires certification of technical personnel involved in the inspection in the chemical and petrochemical industries.

The following is just a small sample of API specifications and codes, which tries to reflect the variety of issues addressed by the institution.

API 2Y	Specification for Steel Plates, Quenched and Tempered for Offshore Structures
API 5L	Specification for Line Pipe
API RP 5L 1	Recommended Practice for Railroad Transportation of Line Pipe
API RP 5L W	Recommended Practice for Railroad Transportation of Line Pipe on Barges and Marine Vessels
API 1104	Welding of Pipeline and Related Facilities
API 6D	Specification for Pipeline Valves (Gate, Plug, Ball and Check Valves)
API 6A	Specification for Wellhead and Christmas Tree Equipment
API RP 5L3	Recommended Practice for Conducting Drop-Weight Tear Tests on Line Pipe
API 650	Welded Steel Tanks for Oil Storage

API also has several maintenance and inspection specifications that are common in use in the petrochemical, oil, and gas industries, some of which are described below.

API 653 (Above-Ground Storage Tanks)

API 653 "Tank Inspection, Repair, Alteration and Reconstruction" is the inspection code for welded or riveted, non-refrigerated, atmospheric pressure, above-ground storage tanks for the petroleum and chemical process industries.

API 510 (Pressure Vessels)

API 510 "Pressure Vessel Inspection Code: Maintenance Inspection, Rating, Repair, and Alteration" is the pressure vessel inspection code for the petroleum and chemical process industries.

API 570 (Pressure Piping)

API 570 "Inspection, Repair, Alteration, and Rerating of In-Service Piping Systems" is the piping inspection code for the petroleum and chemical process industries.

API RP 579 (Fitness for Service)

API RP 579 "Fitness for Service" is a recommended practice (RP). The purpose of the recommended practice is to provide guidance to the methods applicable to assessments that are specific to the type of flaw or damage encountered in refinery and chemical process plant equipment.

API RP 580 (Risk Based Inspection)

API RP 580 "Risk Based Inspection" is a recommended practice (RP). The purpose of the recommended practice is to provide guidance regarding the development of a risk-based inspection program with the methodology presented in a step-by-step manner for users in refinery and chemical process plants.

AMERICAN SOCIETY FOR TESTING MATERIALS (ASTM)

ASTM specifications complement most of the construction specifications. These specifications and codes address several material and testing procedures requirements and guidance. The specifications issued by the American Society of Testing Material (www.astm.org) are organized on the basis of the type of material, and the letter prefixed to the specification number is indicative of the material type, for example letter A is for all ferrous materials, letter B is for all non-ferrous materials, letter C is for Cementations, Ceramic, Concrete and Masonry, letter D is used to indicate specifications related to miscellaneous material such as chemicals, polymers, paints, coatings and their test methods etc. Similarly letter E is used to denote specifications that address miscellaneous subjects, including subjects related to examination and testing of materials.

The following is the short list of some of these groups. The list is only intended to be a general explanation of what is described above.

ASTM A 6   Specification for General Requirements for Rolled Structural Steel Bars, Plates, Shapes, Sheets Pilling.

ASTM A 20   Specification for General Requirements for Steel Plates for Pressure Vessels.

ASTM A 36   Specification for Carbon Structural Steel.

ASTM A 176   Specification for Stainless and Heat-Resisting Chromium Steel Plate, Sheet and Strip.

ASTM A 181   Specification for Carbon Steel Forgings for General Purpose Piping.

ASTM A 351   Standard Specification for Castings, Austenitic, Austenitic-Ferritic (Duplex), for Pressure Containing parts.

ASTM A 370   Standard Test Methods and Definitions for Mechanical Testing of Steel Products.

Similarly the non-ferrous materials list is exhaustive, and a few of them are listed below as examples.

ASTM B   Specification for Hand-Drawn Copper Wire.

ASTM B 80   Specification for Magnesium-Alloy Sand Castings.

ASTM B 159   Specification for Phosphor Bronze Wire.

ASTM B 418   Specification for Cast and Wrought Galvanic Zinc Anodes.

ASTM B 457   Test Method for Measurement of Impedance of Anodic Coating on Aluminum.

ASTM B 491   Specification for Aluminum and Aluminum-Alloy Extruded Round Tubes for General-Purpose Applications.

ASTM B 546   Specification for Electric-Fusion-Welded Ni-Cr-Co-Mo Alloy (UNSN06617), Ni-Fe-Cr-Si Alloy (UNS N08330 and UNS N08332), Ni-Cr-Fe-Al Alloy (UNS N06603), Ni-Cr-Fe Alloy (UNS N06025), and Ni-Cr-Fe-Si Alloy UNS N06045) Pipes.

A sample list of cemetitious, ceramic, concrete and masonry materials is included below.

ASTM C 4   Specification for Clay Drain Pipe.

ASTM C 42   Test Method for Obtaining and Testing Drilled Cores and Sawed Beams of Concrete.

ASTM C 144   Specification for Aggregate for Masonry Mortar.

ASTM C 150   Specification for Portland Cement.

ASTM C 155   Classification of Insulating Firebrick.

ASTM C 173   Test Method for Air Content of Freshly Mixed Concrete by Volumetric Method.

A sample of miscellaneous materials is given below.

ASTM D 20	Test Method for Distillation of Road Tars.
ASTM D 75	Practice for Sampling Aggregates.
ASTM D 98	Specification for Calcium Chloride.
ASTM D 143	Test Method for Small Clear Specimens of Timber.
ASTM D 185	Test Methods for Coarse Particles in Pigments, Pastes, and Paints.
ASTM D 388	Classification of Coals by Rank.
ASTM D 1621	Test Method for Compressive Properties of Rigid Cellular Plastics.
ASTM D 1640	Test Methods for Drying, Curing, or Film Formation of Organic Coating at Room Temperature.

A sample list of miscellaneous subjects is given below.

ASTM E 4	Practices for Force Verification of Testing Machines.
ASTM E 6	Terminology Relating to Methods of Mechanical Testing.
ASTM E 55	Practice for Sampling Wrought Nonferrous Metals and Alloys for Determination of Chemical Composition.
ASTM E 83	Practice for Verification and Classification of Extensometers.
ASTM E 94	Standard Guide for Radiographic Examination.
ASTM E 73	Practice for Static Load Testing of Truss Assemblies.

A sample list of materials for specific applications is given below.

ASTM F 1	Specification for Nickel-Clad and Nickel-Plated Steel Strip for Electron Tubes.
ASTM F 22	Test Method for Hydrophobic Surface Films by the Water-Break Test.
ASTM F 78	Test Method for Calibration of Helium Leak Detectors by Use of Secondary Standards.

A sample list of corrosion, deterioration, and degradation of materials is given below.

ASTM G 1	Practice for Preparing, Cleaning, and Evaluating Corrosion Test Specimen.
ASTM G 5	Reference Test Method for Making Potentiostatic and Potentiodynamic Anode Polarization Measurements.
ASTM G 6	Test Method for Abrasion Resistance of Pipeline Coatings.
ASTM G 11	Test Method for Effects of Outdoor Weathering on Pipeline Coatings.
ASTM G 36	Practice for Performing Stress-Corrosion Cracking Test in a Boiling Magnesium Chloride Solution.

The above lists are just examples of the subjects addressed by each group; in fact the actual list is exhaustive in each case. These specifications are amended, merged, removed or changed from time to time, so the latest up-to-date version must be referenced.

Index